D1403405

ISBN-10: 1974695794
ISBN-13: 978-1974695799

Enzymes, Wizards and Secret Passages

INTUITIVE LESSONS IN BIOCHEMISTRY

Kieran Geoghegan

DEDICATION

To every scientist that I have worked with and learned from.

Thank you.

CONTENTS

Preface

The chemistry of the living cell is beautiful, fascinating and intricate. Exploring it as a student should be a thrilling experience, but the massive weight of facts encountered as we learn biochemistry can be a heavy lift. It adds to the challenge of securing a full understanding of the principles, methods, technologies and ideas behind our certainties and intuitions.

I was a research biochemist for forty years, four decades during which dazzling new discoveries arrived so frequently that they seemed the natural order of things. It took constant reflection and self-questioning to stay up to speed with advances in technology and scientific insight. My reward was to remain grounded in the reality of the science rather than having to take other people's word for things. If I could occasionally make something clearer to a colleague, this was only fair considering how often I pestered better-informed friends to share their wisdom.

The book offers short versions of my personal understanding of selected topics. The accounts are intended to be simple and unintimidating, the sort of friendly chat through fundamentals that can dispel the haze around an unfamiliar topic. Plenty of excellent sources deliver the science in formal terms: *Enzymes, Wizards and Secret Passages* aims for a more intuitive understanding that grounds biochemistry in the fundamental principles of physics and chemistry. It may suit new graduates beginning a first job in industry, some non-science majors, graduate students beginning research, and perhaps the occasional professional scientist or manager meeting an important technology for the first time.

Kieran Geoghegan – 2018

Mystic, Connecticut

Enzymes, Wizards and Secret Passages

1

In science, it can be a mistake to pay too much attention to *words*. Yes, physical phenomena need names so that we can talk about them, but everyday words that scientists use as labels can lead us to misunderstand what is really happening.

Modern physics offers examples. The subatomic entities called particles are not the little pellets that we recognize in the macroscopic world, but exhibit wavelike behavior as well. Entanglement, a word that copes well with a messed-up shoelace, elegantly but incompletely denotes the spooky-action-at-a-distance property behind quantum computing. In fact, physicists generally caution that mathematics is the *only* language that describes realities of this kind. Words convey a pale sense of what is understood, but never its complete meaning.

Living systems are complex, and it took until around the end of the nineteenth century to dispel the idea that a mystical component sets them apart from phenomena governed by the laws of physics and chemistry. Today, most thinkers acknowledge that life needs constant energetic reinforcement against entropic collapse, but confidently view it as compatible with the principles that govern the non-living universe.

Before that awareness was gained, it must have been natural to attribute special properties to living cells. The ability of yeasts to convert sugars into alcohol was first considered unique to live cells, then found to persist in extracts, and eventually associated with

1

molecular entities called enzymes ("in yeast"). The term "activity" is a reasonable one to have pinned onto this ability, but perhaps it has done some damage to our understanding.

Enzymes Then And Now

Together with decades of progress in enzymology, structural biology has given us a direct view of how enzymes function. Today, we classify enzymes by their mechanisms of action, and the digestive proteases trypsin and chymotrypsin are considered siblings even though they will barely touch each other's substrates. But, for the first half of the twentieth century, enzymes had to be named not for what they were but for what they *did*. Under this system, it was difficult to see the molecular relationship between two very similar enzymes that dealt with different substrates. This tradition continues, and enzymes receive trivial names for the reactions that they catalyze, even as underlying similarities of mechanism and structure are used to group them into families, a process in which they also receive more systematic names.

The Fallacy of Activity

Every student of biochemistry is told, correctly, that enzymes are catalysts – they take part in reactions but emerge unchanged when the reactions are over. They don't make reactions happen, they allow them to happen. It can pay to think about this further.

Chemical reactions are thermodynamically favored if their products lie at a lower level of "free energy", or "energy available for work", than their starting materials. A nice analogy to the simplest case is water stored behind a dam, below which lies the natural bed of the dammed river. If the dam fails, the water spontaneously falls to its natural level. Given a route, the favored process takes place on its own. No input of energy is needed.

2

A reaction may be thermodynamically favored, but effectively unable to proceed in the absence of a catalyst because an energy barrier, like a dam, cannot be surpassed under prevailing conditions. Hydrolysis of a peptide bond in a peptide is a good example. It requires strongly acidic conditions and higher temperature to proceed appreciably without catalysis. But when we introduce a proteolytic enzyme with appropriate specificity, the substrate peptide falls apart under mild conditions.

These words are chosen carefully – rephrasing them slightly, the enzyme "allows the substrate peptide to fall apart." We must not imagine the protease as a mad slasher armed with a sword that actively hacks into the bonds of the substrate. Instead, it offers the reaction a viable pathway where no passable route existed before. This is a *passive* function in the very place where we use the term activity.

An analogy might be useful (personally, I can barely think without them). One that serves, but only in part, is the distinction between an arduous route over the high terrain of a mountain and a secret tunnel that allows a much easier passage to the far side (this is the preferred choice of wizards: see Figure 1-1). The enzyme offers the reaction this easier route, a passive form of assistance to its progress.

A weakness in the analogy is that a pathway, unless governed by magic spells, is a rigid and immobile track, whereas enzymes cannot function without being flexible enough to vary their conformations during catalysis. Nevertheless, in strictly energetic terms, it is helpful to see an enzyme as an entity that accepts, channels and steers its substrate through a reaction pathway rather than imposing itself forcefully.

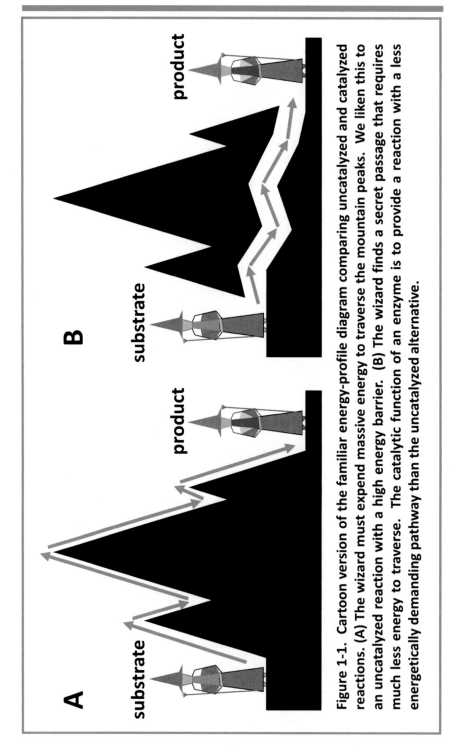

Figure 1-1. Cartoon version of the familiar energy-profile diagram comparing uncatalyzed and catalyzed reactions. (A) The wizard must expend massive energy to traverse the mountain peaks. We liken this to an uncatalyzed reaction with a high energy barrier. (B) The wizard finds a secret passage that requires much less energy to traverse. The catalytic function of an enzyme is to provide a reaction with a less energetically demanding pathway than the uncatalyzed alternative.

Forward and Reverse Reactions

Like the fall of pent-up water, some reactions are thermodynamically strongly favored so that – given an open pathway – they tend strongly to run in one direction. Later in this book we will refer several times to the strongly favored hydrolysis of adenosine triphosphate (ATP), a reaction catalyzed by many different enzymes. Those enzymes have nothing whatever to do with determining whether the reaction is favored. They just provide it with a pathway.

Some reactions are closer to being balanced. An example is the one catalyzed by triosephosphate isomerase, which facilitates the interconversion of glyceraldehyde 3-phosphate and dihydroxyacetone phosphate. Adding the enzyme to a solution of *either one* of these metabolites will result, when equilibrium is reached, in a mixture of the two of them at the ratio determined by the equilibrium constant governing their interconversion (about 20:1 in favor of dihydroxyacetone phosphate).

This shows that the enzyme does not affect where equilibrium lies: it just allows equilibrium to be reached. An enzyme must catalyze the same reaction by the same mechanism in either direction.

Our habit of designating substrates and products for enzyme-catalyzed reactions is a sort of shorthand. The products of the forward reaction are the substrates for the reverse reaction, so our terminology asserts our own, potentially fallible, understanding of the reaction's biological flow and significance.

The levels of substrate and product become constant when equilibrium is reached, but that does not mean that no interconversion takes place between them. The concept of dynamic equilibrium (see Chapter 11), an essential one in biochemistry, means that the reaction pathway is being traversed equally in both directions when equilibrium has been reached. Whether the equivalent forward and reverse rates are fast or slow depends on multiple factors such as concentrations and the catalytic efficiency of the enzyme.

Enzyme Mechanism is Irrelevant to Reaction Equilibrium

Another way to tease apart the principles governing enzyme action is to consider that two enzymes that catalyze the same reaction can have entirely different mechanisms. Mammalian pancreatic carboxypeptidase A is a zinc metalloenzyme in which a divalent zinc ion at the active site helps to activate a water molecule that attacks the most C-terminal peptide bond of a substrate. In contrast, the yeast enzyme carboxypeptidase Y is a serine hydrolase in which an active-site serine residue is activated (has its hydroxyl proton pulled away) so that it can directly attack the substrate peptide bond.

These two carboxypeptidases can catalyze hydrolysis of the same substrate to the same products, but by different mechanisms. They show different kinetic properties, different preferences among a range of substrates, and susceptibility to entirely different inhibitors and inactivators. Crucially, though, the difference in their mechanisms does not affect the equilibrium constant for hydrolysis of a common substrate.

Enzymes are not wizards who can change the course of nature. To the contrary, they are specialists in letting nature take its course.

Decoding Enzymology

2

Separating Catalysis from Substrate Binding

The terminology and equations of enzymology can be difficult to relate to other topics. Some people absorb abstract material more easily than others, but a correct intuitive understanding of the fundamentals is accessible to everyone.

Enzymes do two things. They *bind substrate* and they *effect catalysis*. Without both taking place, activity is impossible. The specificity constant k_{cat}/K_M quantifies the catalytic efficiency of an enzyme for its substrate under specified conditions. This reflects how "good" an enzyme it is. k_{cat}/K_M marries the contributions of the two indispensable functions.

Effecting catalysis

The numerator k_{cat}, known as the catalytic rate constant, describes how fast the enzyme can catalyze the conversion of the enzyme-substrate complex ES to the enzyme-product complex EP. A high value means that the ES complex, when it forms, converts quickly to the complex of enzyme and product(s). Clearly, increasing k_{cat} will make k_{cat}/K_M bigger too, signaling greater catalytic efficiency.

We can think of k_{cat} as a first-order rate constant gauging the propensity of the ES complex to break down productively. This resembles the rate constant governing the decay of a radioisotope (see Figure 2-1). Some radioisotopes have half-lives of fractions of a second while others break down over thousands of years, but these properties are set entirely by nonbiological physical factors.

In the case of enzymes, we can reasonably assume that evolution has favored (and will always select) properties that adapt them well for their biological activities. In this regard, it is not true that "fastest is best" when it comes to enzymes. For example, certain enzymatic functions involved in turning off cellular signals need to be slow so that the message being transmitted has useful duration.

The rate of an enzyme-catalyzed reaction is also called the reaction velocity. V_{max}, the maximum rate achievable in a particular reaction, is reached when the total concentration of enzyme active sites $[E_t]$ is functioning at the maximum rate, and is given by $k_{cat}[E_t]$.

We say "concentration of enzyme active sites" because the term molecule applies even to oligomeric proteins, and many oligomeric enzymes will have more than one active site. The molar concentration of enzyme active sites for a homodimeric enzyme (one with two identical subunits) is double the molar concentration of the enzyme itself, and so on.

Intuitively, we can imagine all available enzyme active sites $[E_t]$ filled with substrate being converted to product at a rate governed by k_{cat}. Now, suppose the enzyme has evolved to permit that rate to be very, very fast. The overall rate at which substrate is converted to products will not now be set by decay of the ES complex, but by the ability of substrate to diffuse to the active site. Therefore, there is a limit to how efficiently enzymes can catalyze reactions. In terms of the specificity constant k_{cat}/K_M, a long-standing estimate places the limit near $10^9 \, M^{-1} \, s^{-1}$, but in some very special cases additional factors raise this by about a factor of ten.

Key learning – be sure to distinguish *rates* from *rate constants*. Rate constants refer to what *can* happen – the intrinsic potential of a chemical step to take place. Rates are what actually happens.

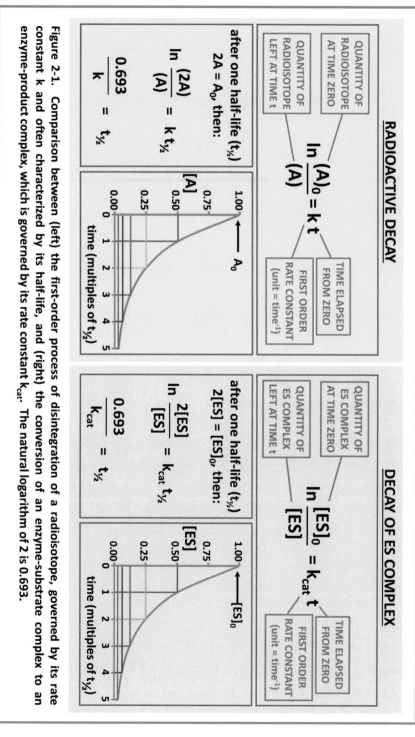

RADIOACTIVE DECAY

QUANTITY OF RADIOISOTOPE AT TIME ZERO

QUANTITY OF RADIOISOTOPE LEFT AT TIME t

$$\ln \frac{(A)_0}{(A)} = k\, t$$

TIME ELAPSED FROM ZERO

FIRST ORDER RATE CONSTANT (unit = time^{-1})

after one half-life ($t_{\frac{1}{2}}$)
$2A = A_0$, then:

$$\ln \frac{(2A)}{(A)} = k\, t_{\frac{1}{2}}$$

$$\frac{0.693}{k} = t_{\frac{1}{2}}$$

$$[A]$$
1.00 — A_0
0.75
0.50
0.25
0.00

time (multiples of $t_{\frac{1}{2}}$)
0 1 2 3 4 5

DECAY OF ES COMPLEX

QUANTITY OF ES COMPLEX AT TIME ZERO

QUANTITY OF ES COMPLEX LEFT AT TIME t

$$\ln \frac{[ES]_0}{[ES]} = k_{cat}\, t$$

TIME ELAPSED FROM ZERO

FIRST ORDER RATE CONSTANT (unit = time^{-1})

after one half-life ($t_{\frac{1}{2}}$)
$2[ES] = [ES]_0$, then:

$$\ln \frac{2[ES]}{[ES]} = k_{cat}\, t_{\frac{1}{2}}$$

$$\frac{0.693}{k_{cat}} = t_{\frac{1}{2}}$$

$$[ES]$$
1.00 — $[ES]_0$
0.75
0.50
0.25
0.00

time (multiples of $t_{\frac{1}{2}}$)
0 1 2 3 4 5

Figure 2-1. Comparison between (left) the first-order process of disintegration of a radioisotope, governed by its rate constant k and often characterized by its half-life, and (right) the conversion of an enzyme-substrate complex to an enzyme-product complex, which is governed by its rate constant k_{cat}. The natural logarithm of 2 is 0.693.

Binding substrate

Meanwhile, the Michaelis constant K_M gauges the enzyme's affinity for the substrate. Like an equilibrium dissociation constant, its units are M (molar). A lower value means higher affinity between enzyme and substrate and a *higher* value for k_{cat}/K_M. This makes sense if we recall that K_M can be taken as the substrate concentration at which the enzyme binds substrate 50% of the time: the lower that concentration is, the greater must be the affinity between enzyme and substrate.

We consider specific binding affinity between molecules in the next chapter. The initial interactions of enzymes and their substrates are binding events, although catalytic function requires enzyme-substrate complexes to undergo conformational shifts between different intermediate states to facilitate chemical change. This topic is deep, so for present purposes let's just accept that increasing affinity between substrate and enzyme tends to contribute to increasing activity.

A prime goal of enzymology is to measure the two elements of activity separately and assess their respective contributions to the enzyme's overall capacity for catalysis. They are phenomena of two totally different kinds, and the essence of science is to recognize them as distinct and measure them separately.

Enzyme Mechanisms

Structural biology has shown that the amino-acid side chains responsible for catalyzing chemical change in the substrate(s) can often be recognized as separate from those involved in binding the substrate(s), even as the two functions of binding and catalysis exist in perfect alignment.

This is illustrated by again comparing trypsin and chymotrypsin, two pancreatic serine proteases that enter the digestive tract to digest proteins in food. They have similar overall folds (Figure 2-2) and catalytic machinery, but distinct substrate specificities. Trypsin catalyzes hydrolysis of peptide bonds following the basic amino acids lysine and arginine, while chymotrypsin cleaves bonds following aromatic and hydrophobic amino acids.

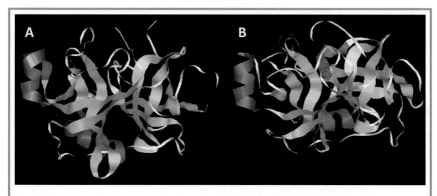

Figure 2-2. Cartoon views of (A) bovine chymotrypsin and (B) bovine trypsin, showing their generally similar overall structures. Figure generated using the program RasMol from Protein Data Bank entries 4CHA (A) and 1S0Q (B). Please see Acknowledgements and Suggestions for further credits.

The substrate-recognition pockets of the two enzymes reflect their respective specificities without fully accounting for them. In trypsin, a negatively charged aspartyl residue is present to interact with the positively charged side chains of lysine and arginine. In chymotrypsin, its place is taken by a neutral serine.

As another illustration, competitive inhibitors often resemble substrates and fit the enzyme's binding specificity, but they are not suitable for catalytic action. The enzyme's ability to bind the inhibitor without catalytic action taking place shows that its binding and catalytic capacities are independent, even though they are intimately mutually dependent when catalysis occurs.

I Love You, You Love Me – Understanding Molecular Binding

3

Specificity and Affinity Underpin Binding

Specific molecular recognition is the basis of biochemical specificity. Enzymes act on particular substrates, multiprotein complexes self-assemble thanks to mutual recognition among their components, and receptors at the cell surface change shape to deliver signals after binding their specific ligands. All of this takes place against a massive background of *non-binding*. Just as a call to your cell phone makes it ring while everybody else's phone stays silent, biochemical signaling is phenomenally specific.

But....Specificity Is a Matter of Degree Rather than Absolute

In biochemistry, as in many matters, it's important to recognize that we often deal with differences of degree, even when our language denotes absolute distinctions. We may say the enzyme binds sugar A but not sugar B – a statement of absolute distinction – while a more accurate statement might use relative terms. Perhaps the enzyme binds sugar A only a thousand times more tightly than sugar B. Under the right conditions, the interaction between sugar B and the enzyme could still be significant. As we already mentioned in Chapter 1, the words that science uses to describe a phenomenon sometimes match inexactly to the reality.

Distinguishing Facts from Terminology

The language that we use to describe molecular binding can give students the impression that a protein molecule, be it an enzyme or a receptor, is a sort of molecular magnet that attracts its small-molecule ligands using special powers of Force. In these mistaken terms, the protein has affinity for the small molecule, and captures it like a giant spacecraft in *Star Wars* opening its hold to swallow a smaller one captured by its tractor beam.

Enzymology is taught in a way that can unintentionally reinforce this view. Michaelis-Menten kinetics is used to introduce the fundamental concepts of saturability, maximum rate and quantifiable enzyme-substrate affinity. We imagine that the enzyme exists at exceedingly low concentration while the substrate concentration is varied. As substrate concentration crosses from the range at which enzyme mostly does not bind it into the range at which it mostly does, velocity measurements reveal the onset of saturation and the Michaelis constant (K_M) is extracted. K_M is interpreted as the substrate concentration at which 50% of the enzyme molecules have substrate bound. The idea might form that the enzyme is an active captor of other molecules and the lowly substrate molecules are its passive prey.

Likewise, when we consider the binding of a ligand to a cell-surface receptor, we commonly imagine that the receptor exists at very low concentration while the ligand is titrated through the range in which binding progressively increases. The reward is to extract the equilibrium dissociation constant, K_d, generally understood as the ligand concentration at which the receptor is 50% occupied by ligand.

K_d serves well as a single-term statement of the receptor's affinity for its ligand. When we say that a tight-binding ligand has "nanomolar affinity", we mean that a single receptor molecule subjected to titration with ligand would be 50% occupied when the ligand concentration is in the nanomolar range.

13

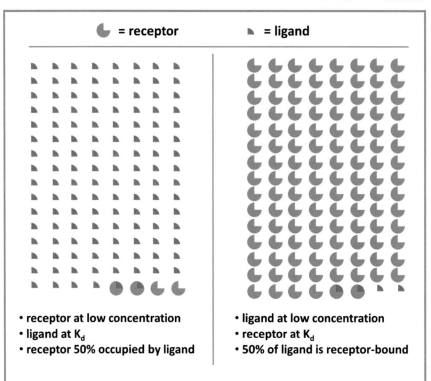

= receptor   = ligand

- receptor at low concentration
- ligand at K$_d$
- receptor 50% occupied by ligand

- ligand at low concentration
- receptor at K$_d$
- 50% of ligand is receptor-bound

Figure 3-1. Cartoon illustration of the meaning of K$_d$. Numbers of molecules shown are not to be taken literally. Key learning: *the affinity of the receptor for the ligand and the affinity of the ligand for the receptor are one and the same thing.* **K$_d$ gauges that affinity, and is a constant. What actually happens depends on the respective concentrations of the two binding partners.**

What Is Left Unsaid Can Be Lacking In What Is Understood

We must remember that the receptor's affinity for the ligand and the ligand's affinity for the receptor *are one and the same thing.* The molecules have affinity for each other. It's not a case of one seeking the other; their tendency to associate is a property that they possess jointly.

If the normal balance of the titration is inverted so that the receptor is at various high concentrations and the ligand is present at exceedingly low concentration, then each ligand molecule will spend 50% of its time on the receptor when the receptor is at K_d (see Figure 3-1).

K_d references *the mutual affinity of the two partners.* It is not a property of one molecule or the other.

Dissociation constants give us useful statements of the affinity between partners. Constants are really constant – they do not vary – but we mainly care about how much binding of one partner to another actually occurs under particular conditions.

If R is receptor, L is ligand and R-L is their noncovalent complex, then:

$$K_d = \frac{[R]\,[L]}{[R\text{-}L]}$$

Rearrangement gives us:

$$\frac{[R\text{-}L]}{[R]} = \frac{[L]}{K_d}$$

Now we see how convenient it is to consider the receptor at a concentration far below K_d. [R-L]/[R] is the receptor's occupancy by ligand, the ratio of receptor complexed with ligand to free receptor. *This is the quantity that we care about,* and we see that it is predictable from the ratio of the ligand concentration [L] to K_d.

Look Beyond Molar Ratios

Take care to understand how binding depends on concentrations relative to K_d, and not just on molar ratios. Let's consider the case in which a ligand L binds to a receptor R with $K_d = 1$ nM.

If [L] = 100 pM and [R] = 1 pM, then [L]/[R] = 100. But [L]/K_d is only 0.1, so only about 10% of R molecules have L bound to them.

If [L] = 10 nM and [R] = 100 pM, *we have the same 100-fold excess of L over R.* But [L]/K_d in this case is 10, and about 90% of R molecules will have L bound to them.

When Molecules Compete for a Receptor

Some very important cases are those in which two molecules represent alternative ligands to the same binding site on the same protein target – examples include competitive inhibition of an enzyme, or competition for a receptor by agonist and antagonist ligands that fill the same site.

Which ligand "wins"? Each has its own specific affinity for the protein that is stated by its own dissociation constant. *The ability of each ligand to compete for the binding site depends on its concentration relative to its own dissociation constant.*

Let's look at a simple example. Two ligands L1 and L2 bind to the same site on a receptor, but L1 has K_d = 5 x 10^{-9} M and L2 has K_d = 1 x 10^{-9} M. These values tell us that L2 has five-fold higher affinity for the receptor than L1.

Now, if we incubate a subnanomolar (<1 nM) concentration of the receptor with 10 nM L1 and 2 nM L2, each ligand is present at twice its own K_d. Under this condition, they should compete on equal terms for the receptor, and equal fractions of receptor should be occupied by L1 and L2.

Immunocapture: A Real-World Example

Despite the common thought process that considers ligand at concentrations in the range of K_d and receptor at much lower concentration, at least one everyday procedure in modern biochemistry reverses this polarity.

Immunoprecipitation is the case in point. An antibody with affinity for a certain target protein is dropped into a biological extract that contains the target molecule. The goal is often to capture the target and see what comes with it in order to learn something about its biological partners. The experiment is simple in principle, but tends to be complicated by nonspecific interactions.

After allowing time for antibody to bind to the target protein, we affinity-capture the antibody to agarose beads and spin out the beads with (we hope) antibody-target complex attached.

We want to capture as much target as we can, preferably all of it, but its concentration is unknown and can be low. Using the analysis presented above, however, we can consider the target as ligand L and the antibody as receptor R. We will know K_d or a reasonable estimate. With a similar rearrangement of the equation to the one presented above, we see that $[R-L]/[L] = [R]/K_d$.

$[R-L]/[L]$ is the result we care about. If it is 10 or more, we will have captured 90% or more of the target L.

Therefore, to maximize immunocapture of the target, we must ensure that the effective concentration of the antibody $[R]$ is ten times higher (or more) than K_d.

Words Again

By now we should realize, if we did not do so earlier, that the designations "ligand" and "receptor" are only terms of convenience that reflect our conventional view of biological systems. This perspective is even baked into our frequent designation of certain cellular proteins as receptors for specific ligands. The practice is only harmful if we fail to realize that we are imposing our chosen frame of reference on the system. Einstein might encourage us to think again.

ATP – How the Cell Extracts Energy From Its Fuel

4

Engines Use Energy Stored in Their Fuel

Most people know how the engine in a traditional car converts the energy stored in gasoline into work. When a compressed vapor of fuel in air is ignited by a spark, the explosion of combustion products drives a piston head along the length of a cylinder. Rods and gears transmit the motion to the wheels, and four or six cylinders working together give us a functioning engine.

In an electric car, charging the battery is like storing water behind a hydroelectric dam, except that water is replaced by electrons raised to a high potential difference, or voltage. To run the motor, we close a circuit that allows current to flow through its armature, and motion is generated according to the discoveries of Michael Faraday. Mechanically coupling it to the wheels allows the car to move.

The mobility of steam locomotives or nuclear-powered ships can be explained in the same way. Controlled release of energy from a primary source – burning fuel or a nuclear fission chain reaction – is harnessed mechanically to achieve work.

The Fuel of the Cell

ATP (adenosine triphosphate: see structure on right) is sometimes called "the fuel of the cell." It's easy, but mistaken, to guess that ATP powers mechanical work in the cell in a way that resembles the examples above – that breakdown of the fuel is *directly* applied to useful work, with ATP hydrolysis somehow resembling a small explosion from which the cell captures the energy. Let's see what really happens.

18

Two things are worth keeping in mind.

First is the type of motion that proteins engage in: typically this is a matter of switching between separate, distinct states. Because these states are not all at the same energy level, the protein needs assistance in order to work its way from a low-energy (or relaxed) state to a higher-energy (or tensed) state. Second is the fact that proteins have to be able to cycle repeatedly between these different states, so that the stimulus allowing the lower-to-higher energy transition must be reversible.

We will argue that the favorable binding of ATP to proteins is the factor that induces them to change shape, and that hydrolysis of ATP is valuable mainly because it allows them to relax back to the initial condition. The fact that hydrolysis of ATP is energetically favored makes it very well suited to this function.

The ATP Molecule

ATP is said to contain "energy-rich" bonds, but what does this mean? Imagine an archer's bow with the string pulled back and an arrow ready to fly. The archer stores potential energy in the bowstring by applying mechanical work to it and holding it in the extended state. While it is pulled taut, the bowstring is "energy-rich" because it is storing potential energy. Releasing it allows it to relax, while the stored energy shoots the arrow.

How Energy Is Stored in ATP

ATP resembles a taut bowstring because it takes work to pack the cluster of mutually repulsive negative charges belonging to the triphosphate into tight proximity. The cell does that work in various ways. Energy captured from the breakdown of nutrients is harnessed through enzymatic reactions or electron-transport chains, and ultimately used to force ATP into existence from its building blocks of ADP (adenosine diphosphate) and inorganic phosphate. This process requires work. When ATP has been made, it represents the cell's primary reserve of refined fuel ready for use.

Contrast the rich diversity of your diet with the singular nature of ATP. Your foods are your fuel in crude form, and ultimately provide the energy that you need to live. For that energy to be usable, though, your metabolism has to process its feedstock so that the energy becomes available to your muscles, brain and other organs *in a form they can use.*

In large part, that form is the energy stored in ATP.

When the archer's bowstring snaps back to its resting length, the stored energy becomes the kinetic energy of the arrow. With ATP, we have to think in terms of thermodynamics and chemistry. It took work to make the ATP molecule, and the reward for loading that small molecule with stored energy is to make its hydrolysis (shown on page 21) a thermodynamically favored process.

This mode of hydrolysis – cleaving the terminal phosphoester bond while introducing the elements of H_2O – is most often used in the cell to split ATP, producing ADP and inorganic phosphate. Sufficient *activation energy* has to be provided. By this we mean enough of an energetic investment to get the process started, after which it goes to completion because it is thermodynamically favored.

Calorimetry can gauge the energy balance of this process by measuring the heat that it releases into solution. It shows that hydrolysis of the terminal gamma-phosphate of ATP yields from -30 to -50 kJ/mol depending on the conditions tested (e.g. whether magnesium ion is present or not). The quantity is negative because the reaction is exergonic, or accompanied by a liberation of energy.

In more relatable terms, hydrolysis of 0.5 g of ATP releases enough heat to raise the temperature of 1 g of water by around 10 °C. Clearly, the process is energetically favorable like the combustion of gasoline, the downhill flow of water, the fall of electrons from high potential, or spontaneous fission of a metastable nucleus.

The Function of ATP Used as a Fuel Is to Bind to Proteins

In biochemistry, we have to identify the mechanism by which the favored process of ATP hydrolysis is coupled to useful work that requires energy. The answer hides in plain sight.

Proteins are large, flexible molecules that often bind smaller molecules called ligands. ATP is frequently a ligand to proteins.

Now, most proteins are flexible – many of them, including enzymes, solute carriers and chaperones, must shift their shapes to function – but their different conformations do not exist at the same level of energy. Consider a protein that exists in two distinct states, A and B, and that it takes energy (work) to shift it from state A to state B. On the other hand, unliganded B-state protein relaxes spontaneously to the A state and no major energy barrier prevents it from doing so. The protein will not populate the B state much unless nature has found a way to facilitate the A-to-B transition.

How would energy be supplied to make that possible? Suppose the B state of the protein can bind ATP, and the binding of ATP is so energetically favorable that the protein will adopt the B state in order to make it happen. Now we have used ATP to make an energy-requiring process take place. Figure 4-1 develops this story a little further, using the example of a protein-protein interaction.

Up to this point, ATP has not been hydrolyzed. Did we get something for nothing? No, because we are not finished. Changing the shape of our protein from A to B was useful, but now B state protein is bound to ATP. ATP isn't going to dissociate from the protein because its binding to the B state was a favored process.

We escape the trap and return to the starting point by hydrolyzing ATP to ADP and inorganic phosphate, two products that bind less tightly to the protein than ATP. Their complex with the protein falls apart, the protein relaxes to the A state, and is ready to repeat the useful behavior of binding ATP and undergoing the A to B transition.

Many proteins have evolved to be able to hydrolyze bound ATP and to have lesser affinity for the products of its hydrolysis, which proceed to dissociate. This activity leaves our protein able to relax back to the A state. Useful work has been done, and the hydrolysis of one molecule of ATP has been used to pay for it – but hydrolysis itself was not the step at which a favored (exergonic) process was

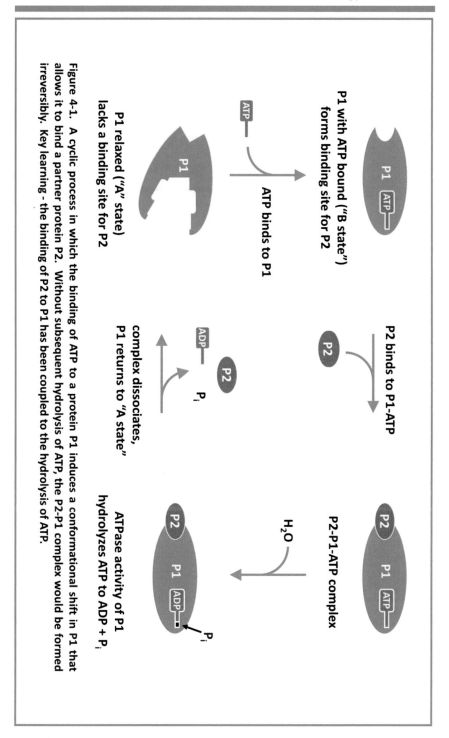

P1 with ATP bound ("B state") forms binding site for P2

P2 binds to P1-ATP

P2-P1-ATP complex

ATP binds to P1

ATPase activity of P1 hydrolyzes ATP to ADP + P_i

P1 relaxed ("A" state) lacks a binding site for P2

complex dissociates, P1 returns to "A state"

Figure 4-1. A cyclic process in which the binding of ATP to a protein P1 induces a conformational shift in P1 that allows it to bind a partner protein P2. Without subsequent hydrolysis of ATP, the P2-P1 complex would be formed irreversibly. Key learning - the binding of P2 to P1 has been coupled to the hydrolysis of ATP.

coupled to an energy-requiring one. That was the step at which ATP bound to the protein. *Hydrolysis of ATP was just a necessary additional step to get the system back to the starting point.*

This is the key learning. The cell can use ATP to do useful work because proteins have evolved to change shape when they bind to it. Its hydrolysis – an energetically favored process – is just a way of disposing of it when it has served its purpose.

In a single sentence, *the function of ATP used as a fuel for mechanical work is to be bound by proteins.*

ATP has other very important functions in the cell, both in signaling functions such as protein phosphorylation and as an intermediate activator of essential chemical processes (see Chapter 8).

Example of ATP Binding Driving Mechanical Work – P2X4 Channel

Purinergic receptors are transmembrane channel proteins that open when they bind extracellular ATP, and then hydrolyze ATP to return to the closed state. They are formed from three identical peptide chains that combine by twisting into a combination with a three-fold axis of symmetry running longitudinally down through the transmembrane pore that they form in the open state. With three chains present, there are three binding sites for ATP that exist at the boundaries between pairs of chains.

Figure 4-2 shows the conformations of the channel in (A) the absence and (B) the presence of ATP. If the channel could not hydrolyze ATP, it would be locked into the open form. Evolution presumably has favored a rate of hydrolysis that provides a useful degree of function. Again, the key point is that the channel opens as a result of ATP <u>binding</u>. Its binding is so energetically favored that the protein changed shape to permit it. Subsequent hydrolysis of ATP allows a cycle of opening and closing to occur.

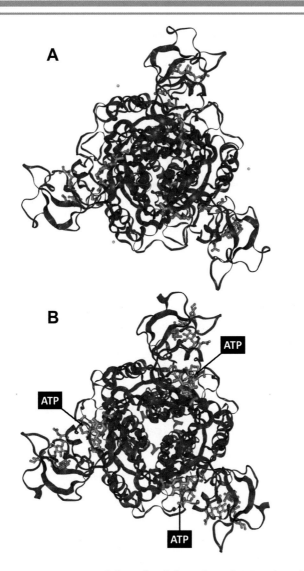

Figure 4-2. X-Ray structures of the zebrafish purinergic P2X4 ion channel in (A) the closed form, and (B) the open form induced by the binding of ATP. Three identical polypeptide chains form the channel, viewed here from the extracellular side. ATP molecules bound to the open form are indicated by arrows. Models were generated using NGL Viewer (AS Rose and PW Hildebrand (2015) Nucleic Acids Res. 43 (W1): W576-W579) from PDB accessions 4DW0 and 4DW1, respectively. Please see Acknowledgments and Suggestions for additional relevant citations.

pH: the Parameter with a Difference

5

When we learn about standard units (meter, kilogram, second, etc.) and their progeny (newton, joule, pascal, etc.), part of the message is that they are universal. Their meaning is exactly the same in any corner of the universe, or at least where space-time as we know it is encountered. (For black holes, call Dr. Hawking).

When we learn biochemistry, we soon start hearing about values of pH. Its definition, $pH = -\log [H^+]$, is not exactly user-friendly, and we also are told that "protons" or "hydrogen ions" are its currency. The proton is a subatomic particle – physicists send it whizzing around the Large Hadron Collider – so this may not clarify our understanding either. Let's define what we are really talking about.

The centrality of *water* to biochemistry makes pH a pivotal parameter to biochemists. The kilogram, for example, can be applied without complications to any particular substance: whether we measure the mass of steel, kryptonite or jelly beans, the unit keeps its austere independence. But, *for biochemists at least, the meaning and scale of pH relates almost entirely to properties of water and aqueous solutions.*

Water molecules exist in several states. The neutral form is H_2O, but it has some ability to dissociate (fall apart) in a way that donates H^+ to another water molecule, converting it into H_3O^+, a hydronium ion. The same process creates OH^-, an hydroxide ion.

$$2H_2O \quad = \quad H_3O^+ \ + \ OH^-$$

Pure water exists at a molar concentration of 55 M. In pure water at 25 °C, dissociation occurs to the extent of creating 1×10^{-7} M concentrations of both H_3O^+ and OH^-. On average, therefore, about two in every billion water molecules in pure water at 25 °C dissociate in this way.

As a shorthand, the concentration of $[H_3O^+]$ is usually referred to as $[H^+]$, and from this comes the misleading reference to protons. It's important to grasp the real nature of what is present. Having done so, we have little choice but to use the convention of referring to $[H^+]$ when we mean $[H_3O^+]$, and also to go along with the use of the term proton in place of the more correct hydronium ion for H_3O^+.

Now for that definition, which can require some thought. pH is defined as the additive inverse of $\log_{10}$ of the concentration of $[H^+]$, that is, $-\log [H^+]$. Therefore, the pH of pure water at 25 °C is equal to 7 because $-\log_{10} [1 \times 10^{-7}] = 7$. (More strictly, it is *activity* or "effective concentration" rather than simple concentration that is reckoned. Activity is dimensionless, and so pH has no unit).

Brønsted-Lowry acids are compounds (A-H) that tend to donate H^+ to water, increasing $[H^+]$ and decreasing pH. Solutions containing an acid of this type generally have pH < 7.

$$H_2O \;+\; A\text{-}H \;\;=\;\; H_3O^+ \;+\; A^-$$

Brønsted-Lowry bases are compounds (B:) that tend to remove H^+ from water and keep it to themselves.

$$H_2O \;+\; B: \;\;=\;\; B:H^+ \;+\; OH^-$$

Then to maintain electrical neutrality:

$$H_3O^+ \;+\; OH^- \;\;=\;\; 2H_2O$$

Here [H^+] decreases and $-\log_{10}$ [H^+] increases from their respective values in pure water, causing solutions of a base to have pH >7.

It might not yet be clear why this is so, so let's take a moment to play with the definition of pH.

When [H^+] decreases from its pure-water, "neutral" activity of 1×10^{-7} (a concentration of 1×10^{-7} M) , the value of the exponent in this expression becomes more negative. For example, a ten-fold drop in [H^+] gives activity of 1×10^{-8}. Taking "minus $\log_{10}$" of this value gives us pH 8.0. Can you see why a hundred-fold fall in [H^+] from its neutral value gives us pH 9.0?

Strong acids have a very high tendency to transfer a proton to water. Examples are hydrochloric acid (HCl) and nitric acid (HNO_3). The pH of a 0.001 M solution of HCl is 3.0, for example, meaning that $-\log_{10}$ [H^+] in such a solution is 3.0, and [H^+] is 0.001 M. In other words, 100% of the dissolved acid molecules (pure HCl is a gas) have dissociated to give H_3O^+ and chloride ion, Cl^-.

Compare this solution with pure water.

- In 0.001 M HCl (pH 3.0), [H^+] = 1×10^{-3} M
- In pure water (pH 7.0), [H^+] = 1×10^{-7} M
- pH 3 is therefore 10,000 times more acidic than pH 7.

Key learning: Because pH is a logarithmic scale, *each unit represents a ten-fold difference in the hydrogen ion concentration*. It's essential to grasp this.

An example of a weaker acid is acetic acid, CH_3COOH. Like HCl, it also tends to give up a proton to water, but its ability to do so is not as strong. As [H^+] increases and pH falls, protons increasingly stay with the acid rather than being transferred to water. At pH 4.7, we reach the point at which this happens in 50% of acetic acid molecules. This value of pH is called the pK_a, or ionization constant.

The Henderson-Hasselbalch equation shows how pH is governed by the pK_a of an acid A-H and its concentration:

$$pH \quad = \quad pK_a \quad + \quad \log_{10} [A^-]/[A-H]$$

When [base] = [acid]   (i.e. when $[A^-]$ = [A-H]):

- $[A^-]$/[A-H] equals one
- $\log_{10} [A^-]/[A-H]$ equals zero
- and the pH is equal to the pK_a of the acid.

Let's look at how $[H^+]$ changes across the pH scale (Figure 5-1). The graph shows how it changes ten-fold with each unit change in the pH. It's true that this is something of a back-to-front statement, because pH is defined in terms of $[H^+]$ – but some biochemists are presented with pH as the primary fact and have to rationalize its physical meaning on their own.

In particular, the idea that pH 7 somehow represents a dividing line between acidic and basic solutions needs to be handled with care. It's important to realize that solutions with pH >7 still have a non-zero concentration of protonated water. That concentration, however, is less than the concentration present in pure water.

Consider what happens when we add a basic compound like ammonia (NH_3) to water. This compound is able to acquire protons from water, itself becoming NH_4^+ and creating equivalent amounts of OH^-.

$$H_2O + NH_3 \quad = \quad NH_4^+ + OH^-$$

Then, as before, to maintain electrical neutrality:

$$H_3O^+ + OH^- \quad = \quad 2H_2O$$

$[H_3O^+]$ is reduced, and pH is raised.

And, as our last look into the chemicals cabinet, a tricky additional case is that of boric acid, $B(OH)_3$. In solution, this compound can abstract OH^- from water, raising the concentration of H_3O^+ and thereby earning itself the title of an acid. In contrast to the Brønsted-Lowry proton-donating acids that we discussed above, boric acid is a Lewis acid (an electron pair acceptor). Its pK_a is near pH 9.2. This is the value of pH at which the free acid $B(OH)_3$ and the borate ion $B(OH)_4^-$ are in equilibrium.

Because we are used to units that we can apply to any relevant case in the universe (the kilogram, etc.), the intimate connection between the pH scale and a single, specific substance – water – requires us to make a small mental adjustment. pH 7.0, the point of neutrality in pH, is a value given to us by a particular molecular property of water. The nearest parallel may be with the Celsius (centigrade) scale of temperature, where key reference points are the freezing and boiling points of water (again) under defined conditions. With pH, we derive a scale by counting changes by factors of ten in the hydronium ion concentration to either side of the neutral value. For the Celsius scale, in contrast, we divide the territory between the two reference points into one hundred parts called degrees, and also extrapolate the scale beyond the reference range to lower and higher values.

pH is a profoundly important parameter in biochemistry. Beyond the everyday use of buffer solutions in laboratory work, the management of pH within compartments of the living cell is a matter of great significance. Appreciation of the logarithmic nature of the scale and the true extent of differences between separate values is a topic well worth some expense of mental effort.

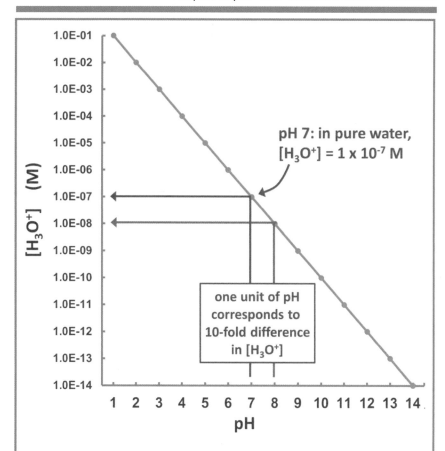

Figure 5-1. Logarithmic nature of the pH scale, and its absolute dependence on physical properties of a particular substance (water). Note how the quantity $[H^+]$, more correctly called $[H_3O^+]$, changes 10-fold with each unit of pH. Note also that at pH >7, generally called basic pH, $[H^+]$ has not gone to zero but simply diminishes by a further factor of ten with each unit increase in pH. Therefore, chemical reactions that depend on protonation of a reagent or intermediate can still occur at pH >7.

Please Don't Mention It (The Absorbance Unit)

6

Measuring Concentrations

We often need to know the concentration of a solution of protein, nucleic acid or other biochemical. One of the best and quickest ways to find this out relies on the fact that many of these molecules absorb light in highly characteristic ways. By relating the **absorbance** of their solutions to standard values, we can estimate their concentrations in molar or mass-based terms.

Absorbance Measurements (Single-Wavelength)

Beer's Law, also called the Beer-Lambert Law, states that the quantity of light absorbed by a substance dissolved in a light-transmitting solvent is directly proportional to its concentration and the distance through the solution traversed by the light.

We measure absorbance using a **spectrophotometer**. In a "single-beam" instrument (Figure 6-1), ultraviolet (uv) or visible light passes through a monochromator, and light of the selected wavelength advances through a cuvette containing the sample. Beyond this is a detector that registers the intensity of the light.

The cuvette must be made of a material that allows the light to pass through. Glass and plastics block uv light, but cuvettes made of quartz allow it through. This is why we must use quartz whenever we measure absorbance at a uv wavelength (e.g. 260 nm, often used for nucleic acids). We can use glass or plastic when the wavelength is in the visible range (from about 350 nm up to about 700 nm, beyond which lies the infrared).

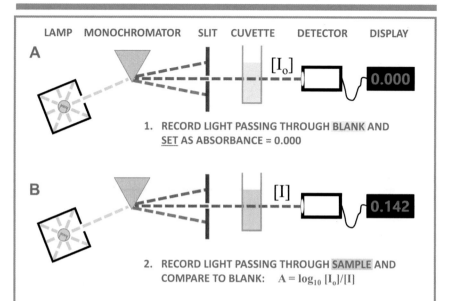

LAMP   MONOCHROMATOR   SLIT   CUVETTE   DETECTOR   DISPLAY

A

$[I_o]$

0.000

1. RECORD LIGHT PASSING THROUGH BLANK AND
   SET AS ABSORBANCE = 0.000

B

$[I]$

0.142

2. RECORD LIGHT PASSING THROUGH SAMPLE AND
   COMPARE TO BLANK: $A = \log_{10} [I_o]/[I]$

Figure 6-1. Key steps in reading absorbance with a single-beam spectrophotometer. **(A) Setting the blank value.** The intensity $[I_o]$ of light that passes through the blank is a reference value that we arbitrarily set to zero absorbance. Subsequent samples are expected to allow an equal or lower intensity of light to pass. **(B) Reading absorbance.** The sample is inserted and the intensity $[I]$ of light passing through it is recorded. Absorbance = $\log_{10} [I_o]/[I]$.

Absorbance measurements gauge the extent to which the presence of a sample <u>reduces</u> the amount of light reaching the detector relative to a baseline value called the blank.

We first register the level of light passed to the detector with a blank sample in the light path. The blank consists of the cuvette filled with the buffer or other solvent in which the sample is made up. We command the instrument to save the blank value, and all later readings are related to it.

Be sure that the blank itself allows light to reach the detector! If in doubt, begin by setting a blank with water (or no cuvette at all) and take a reading against it with your intended experimental blank.

The blank represents the *maximum* level of light that the detector will ever receive in the course of the experiment. We assign it a value of zero, because we are going to determine how much light is *prevented* from reaching the detector by each unknown sample.

Next, we replace the blank sample with one that is equivalent except for the possible presence of dissolved components that absorb light at the selected wavelength. If we have many samples, such as fractions from chromatography, we read them one after another.

As instructed, the instrument remembers how much light passed through the blank. Subsequent samples either match the blank or, compared with the blank, allow less light to reach the detector. Those values of absorbance are zero in the first case, and positive numbers in the second (see Figure 6-1).

A reading below zero means we have done something wrong. For example, the blank may have contained a light-absorbing component that is absent from the sample now being read. This undercuts our attempted measurement. We must go back to the start and eliminate the discrepancy.

Let's Think More Quantitatively

As the concentration of light-absorbing solute increases, the fraction of incident light absorbed by the sample also increases. The quantity called *absorbance* is defined as the logarithm to base 10 of the ratio of the intensity of light transmitted through the blank to the intensity of light transmitted through the sample.

$$A = \log_{10} [I_o]/[I]$$

Consider the mathematical properties of this simple equation.

First, we see that it aligns with the description given above. In the case where the sample gives a reading [I] equal to the blank reading [I_0], the ratio [I_0]/[I] will equal one. Then $\log_{10}$ [1] equals zero, matching our expectation. Absorbance of zero means that the sample absorbed the same amount of light as the blank.

Next, if the sample absorbs more light than the blank, less light reaches the detector than when the blank was read. Therefore, [I_0]/[I] is greater than one, and $\log_{10}$ [I_0]/[I] – the absorbance – is also greater than one.

Suppose the sample blocks 90% of the light that passes the blank; in other words, the blank lets through ten times more light than the sample. Now [I_0]/[I] = 10, and $\log_{10}$ [I_0]/[I] = 1.0.

Then, if the sample blocks 99% of the incident light, [I_0]/[I] = 100, and $\log_{10}$ [I_0]/[I] = 2.0.

This is a key learning. When you read an absorbance of 2, the instrument is telling you that only 1% of the light level recorded from the blank can make it through the sample. Be reasonable about what you expect the instrument to do. Accurate detection of the difference between 99.0% of the light being blocked and 99.9% is effectively not possible. Therefore, on-scale absorbance readings can generally only be obtained in the range between zero and 2.

If samples in a set or series all give readings in the high end of the range and you need to know the difference between them, they should be diluted to make more accurate readings possible.

Another tactic is to shorten the path length of the cuvette employed: halving the path length will halve the absorbance. A new blank reading must be taken with the narrower cuvette.

As we have seen, absorbance is a ratio. *There is no such thing as an absorbance unit.* Some top instrument makers persist in trying to be helpful by showing absorbance in AU (absorbance units), but this is misleading and ultimately not helpful.

Reference Values, and Extracting Concentrations

We noted that experimental readings of absorbance come to a maximum value of 2, at which point the sample blocks 99% of the light able to pass through the blank. Rather than recording higher readings, we should dilute the sample or reduce the cross-section of sample (width of the cuvette) to get a reading from the useful part of the range.

For reference purposes, though, it is perfectly OK to measure or calculate the *molar absorptivity*, also called the *extinction coefficient*, for a particular analyte at a named wavelength.

If a protein, for example, has a molar absorptivity $\varepsilon_{1\,cm}^{280}$ = 6.4 x 10^4 M^{-1} cm^{-1} (the value can be calculated from its amino-acid composition), this means that a one molar solution of the protein would, theoretically, have an absorbance of 64,000 per cm of light path (cuvette width) at 280 nm. Comparison of on-scale experimental absorbance measurements of solutions of this protein to the standard value allows us to calculate the protein concentration in those solutions.

The molar unit (M) refers to moles per liter, abbreviated mol L^{-1}, so the unit of molar absorptivity (M^{-1} cm^{-1}) can also be stated as L mol^{-1} cm^{-1}. This is formally correct, but M^{-1} cm^{-1} conveys the meaning of the unit much more successfully.

Molar or mass units are interchangeable by calculation when the protein's molecular weight is known.

Absorbance Measurements (Multi-Wavelength)

A wealth of information can be gained by measuring absorbance across a spectral range rather than at a single wavelength. A scan of this kind is called an absorption spectrum. Briefly, the absorbance at small intervals across a series of different wavelengths is measured, which requires a blank value at each wavelength followed by many wavelength-specific readings for the sample.

In everyday work, however, you will often collect spectra using an instrument that delivers a multiwavelength band of light through

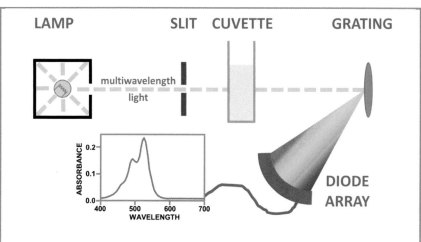

1. AT EACH ELEMENT OF DIODE ARRAY, RECORD LIGHT PASSING THROUGH BLANK AND <u>SET</u> AS ABS = 0.000
2. REPLACE BLANK WITH SAMPLE AND READ: RECORD ABSORBANCE AT EACH ELEMENT OF DIODE ARRAY
3. FROM ARRAY OF READINGS, CONSTRUCT SPECTRUM

Figure 6-2. Schematic view of the design and use of a diode-array spectrophotometer. Because this instrument passes the entire beam of light through the sample before dispersing it into a spectrum, spectra can be collected in 1-2 seconds. These instruments are excellent for routine work.

the sample, then splits it up using a diffraction grating to cover a spread of wavelengths. A bank of photodiodes (1024 in one workhorse instrument) is then used to record the respective light intensities at each small section of the spectrum (Figure 6-2).

For research applications requiring the most precise absorption spectroscopy, the type of instrument used resembles the one shown in Figure 6-1 except that the monochromator can be programmed to scan across a selected part of the light spectrum emitted by the lamp. This allows a spectrum to be collected that effectively consists of a series of single-wavelength readings that covers the range of interest.

Why Is Fluorometry So Sensitive?

7

Comparing Fluorescence with Absorbance

When we considered absorbance, we saw that the baseline measurement (or blank) represents the *maximum* level of light that ever reaches the detector in our instrument. The samples that we compare to the blank absorb more or less of this light, and the measurements max out when absorbance prevents 99% or more of the reference level of light from reaching the detector.

Fluorescence-based measurements are superficially similar, in that light passes into and out of a sample and is registered by a detector (Figure 7-1). Even so, there are fundamental differences between the two methods, some of which are shown in Table 7-1.

With some oversimplification, we can say that fluorescence measurements are generally more sensitive than absorbance measurements. More accurately, we can say that fluorescence-based methods often allow us to detect molecules present at concentrations below 1 μM, while this is not usually convenient when using absorbance. But the differences go far beyond this.

What Happens When Fluorescence Occurs?

The bonds that hold organic molecules together can vibrate like guitar strings. Adding energy by absorption of a photon raises the energy level of this vibration, bringing the bond to an *excited state*. The excited state usually has a short lifetime in the nanosecond range, after which it decays back to the lower-energy *ground state*. To allow this, the molecule has to dispose of the excess energy.

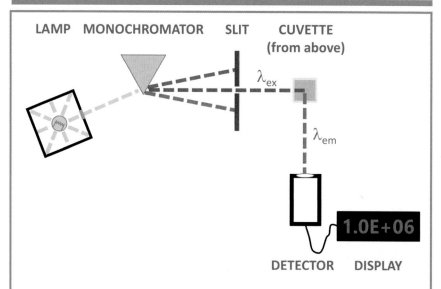

LAMP MONOCHROMATOR SLIT CUVETTE
(from above)

λ_{ex}

λ_{em}

1.0E+06

DETECTOR DISPLAY

Figure 7-1. Relative arrangement of the light source, monochromator, sample cuvette and detector in a standard spectrofluorometer. Key point: the light collected by the detector is emitted from the sample *against a dark background*. Consequences are (i) the potential for high sensitivity, and (ii) the lack of an absolute scale for the measurement. The number of photons emitted by the sample depends on the amount of light of excitation sent into it and many other characteristics of the system.

It can do this in more than one way. One is by emitting a photon of light, which we can detect and measure. We call this process fluorescence. The wavelength of emitted light (λ_{em} in Figure 7-1) is always higher than that of the light used to create the excited state (λ_{ex}), for reasons discussed below.

Proteins are made naturally fluorescent in the near-uv range by the presence of the indole side chain of tryptophan, a property that has been useful in studies of their folding. Even more important in biochemistry are synthetic fluorescent dyes, of which there are very many: Figure 7-2 shows two examples. To avoid optical interference from intrinsic components of cells and tissues (unwanted absorbance, for example), dyes that can be excited and

Table 7-1. A Comparison between Major Features of Absorption and Fluorescence Spectroscopies.

Feature	Absorbance	Fluorescence
Wavelength	Single- or multiwavelength light passes through sample: at each wavelength, absorption is compared to absorption by blank	Input light excites targeted fluorophore; excited fluorophore emits light at higher wavelength (lower energy)
Number of chromophores	May be a mixture, but absorbance at each wavelength interval is basis of measurement	Energy relay between two chromophores can measure molecular interactions or separations
Light path	Straight through sample to detector	Emitted light is detected against a dark background, allowing extreme sensitivity
Scale	Absorbance is $\log_{10}$ of ratio of intensity passing through blank to intensity passing through sample: value of 2.0 indicates 99% absorption	Scale is arbitrary; specific to each instrument and changes with instrument settings (e.g. amount of light admitted to sample)
Common error	Attempting to use readings that represent near-total absorbance of incident light	Excessive absorbance of incident and/or emitted light by sample ("inner filter effect")

yield emission in the visible part of the spectrum have been preferred for several decades. More recently, dyes that function in the near-infrared have become important in studies that require light to penetrate into tissues.

The physical properties of dyes, such as water- or lipid-solubility and their adaptability to chemical coupling, are also important in making them suitable for a wide range of studies.

Many ready-to-use reactive dye reagents are commercially available, and includes the examples shown in Figure 7-2. Fluorescein-related and rhodamine-related dyes are all based on the fused triple-ring system of xanthene, with chemical substitution leading to a rich catalog of variants.

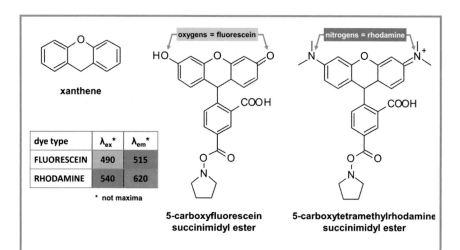

Figure 7-2. Two classic fluorescent dyes shown as reactive forms suitable for modifying amino groups in proteins. Despite a wide range of names, some proprietary, many fluorescent dyes are substituted versions of either fluorescein or rhodamine with slightly different properties from the parents. Fluoresceins have oxygen in the indicated positions, but rhodamines have nitrogen. The excitation (λ_{ex}) and emission (λ_{em}) wavelengths given in the mini-table are wavelengths typically used with these dyes.

Frequently asked questions about fluorescence in biochemistry

Why is the excitation wavelength always lower than the emitted wavelength?

We learn in physics class that $E=hv$; that is, the energy E of electromagnetic radiation (including light) is equal to Planck's constant h multiplied by the frequency v (the Greek letter nu). Higher frequency, or shorter wavelength, means higher energy.

If we sent light of a certain wavelength into our experiment and light of a shorter wavelength (higher energy) came out, we would have created energy from nothing, which is not in our power. Instead, we know that all physical processes dissipate some energy as entropy (disorder or heat).

This should make it easy to understand why the excitation wavelength used in your fluorescence measurement will always be lower than the wavelength of light detected as readout.

I keep hearing about something called FRET. What is it?

FRET is usually read as fluorescence resonance energy transfer, although the first initial "F" can also stand for Förster, after the discoverer. In yet another reading, the "R" can stand for "radiationless", which helps to convey a sense of the process.

FRET is a route of excited state discharge that is an alternative to the emission of a photon. If an excited fluorophore is sufficiently close to an appropriate partner, the excess energy of its excited state can be transferred by a nonradiative process to the partner. One key property of a pair of partners is that the emission spectrum of the "donor" – the initially excited group – must overlap the absorption spectrum of the receptor.

The physical explanation of FRET is complex, and pertains to energy transfer from one oscillating dipole to another. For biochemists, a reasonable analogy is the transfer of energy from a vibrating tuning fork to a frequency-matched partner, although no intermediate medium such as air or water contributes to FRET.

A critical aspect of the process is acute distance-dependence. Resonance energy transfer is inversely proportional to the *sixth* power of the distance separating the donor and acceptor. Therefore, they must be in close proximity for energy transfer to occur, and occurrence (or not) of the process can be taken as a gauge of the distance between the two groups.

The donor group must be potentially fluorescent for FRET to be possible. If the acceptor is also a fluorescent group, the beautiful possibility develops of an energy-transfer relay whereby light that excites the donor leads to fluorescence emitted by the acceptor. Systems that avail of this design can be phenomenally sensitive, as emitted light comes from a background completely free of any scattered light at its particular wavelength.

The acceptor group need not be fluorescent for FRET to work. It can serve simply to quench the native fluorescence of the donor. In this case, fluorescence of the donor group is monitored, and its quenching or dequenching signals change in the system.

Is FRET the only way in which fluorescence can be quenched?

No. Again, think of the bonds in a fluorescent dye as resembling the strings of a guitar. If the strings cannot vibrate freely, the instrument will give no music. A fluorescent group can lose its fluorescence due to contact or static quenching, which occurs – as the name implies – when it is in contact with a group for which it has affinity.

Fluorescent dyes, especially rhodamines, were massively important in the development of lasers, and it was noted that they have moderately high self-affinity. At high concentration in aqueous solution, they tend to form a nonfluorescent dimer with a characteristically shifted absorption spectrum. In general, if there is concern that a fluorescent group is quenched by static quenching rather than hoped-for FRET, it is helpful to examine its absorption spectrum for signs of a shift compared to the free form of the dye.

What is time-resolved fluorescence?

Excellent question! The name of the phenomenon is self-explanatory – fluorescence is followed *over a period of time* that starts with an excitation event, rather than in continuous mode. The process is flash-wait-record rather than continuous excitation along with continuous detection.

We mentioned that the excited states of typical organic fluors have lifetimes measured in tens of nanoseconds, which would make it very challenging to follow the time course of their decay. However, chelated lanthanide ions such as Tb^{3+} and Eu^{3+} decay from their excited states over very much longer times that exceed 10 μs, and this opens up valuable experimental opportunities. It means that we can give a system a flash of excitation, allow any immediate random scattering of light to dissipate, and then monitor the decay of the induced fluorescence under conditions where the system is completely dark apart from the emitted light.

Time-resolved FRET measurements are possible when a lanthanide fluorophore is part of an energy-transfer donor-acceptor pair.

What are green fluorescent protein (GFP) and its relatives? These proteins originate in marine jellyfish, and contain fluorescent chromophores that self-assemble from the protein itself without additional groups being required. As a result, recombinant expression of GFP and its variants as fusions with cellular proteins has become a massively important tool in cell biology. Fluorescence microscopy is used to study the localization of these proteins within the cell, while FRET between proteins fitted with complementary fluorophores is used to assess the extent and timing of association between them.

And what are fluorescence polarization measurements for?

As we discussed, the excited states of organic fluorophores are short-lived, with half-life often in the range of 1-10 ns. Excitation of a fluor often depends on its orientation, so exciting a dye with polarized light will create fluorescent emission that is itself polarized to some extent. That extent will be related to how rapidly the molecule is tumbling in solution due to Brownian motion.

As Figure 7-3 shows, the rate of tumbling changes if a small molecule containing the fluorescent group becomes bound to a macromolecule, such as an antibody or other protein. Fluorescence polarization-based assays for molecular interactions are important in biochemistry.

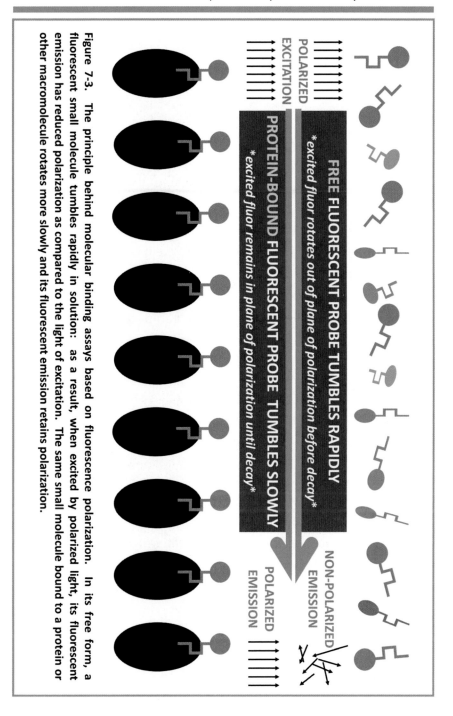

POLARIZED EXCITATION

FREE FLUORESCENT PROBE TUMBLES RAPIDLY

excited fluor rotates out of plane of polarization before decay

PROTEIN-BOUND FLUORESCENT PROBE TUMBLES SLOWLY

excited fluor remains in plane of polarization until decay

NON-POLARIZED EMISSION

POLARIZED EMISSION

Figure 7-3. The principle behind molecular binding assays based on fluorescence polarization. In its free form, a fluorescent small molecule tumbles rapidly in solution: as a result, when excited by polarized light, its fluorescent emission has reduced polarization as compared to the light of excitation. The same small molecule bound to a protein or other macromolecule rotates more slowly and its fluorescent emission retains polarization.

Chemical Activation and Molecular Handles

Stuart Kauffman, a distinguished scholar of the origin and energetics of life, wrote that life "exists at the edge of chaos." A rock is not much good at being alive, he would say: it specializes in remaining unchanged. Nor is a plasma of ionized gas up to the job: it lacks stability. Life is simultaneously stable enough to be visibly continuous but unsettled enough to grow and change. This middle way that life must choose is reflected in biochemistry. Major cellular components cannot freely engage in uncontrolled chemical activity, but there must be ways to equip them with enough reactivity to support life's dynamic nature.

Chapter 4 described how the cell uses ATP to power mechanical work such as changing the conformation of a protein. The cell also has a colossal amount of chemical work to do, especially the task of assembling complex molecules from small building blocks. Here again, ATP is the main source of the needed energy.

Generally this is done by drawing energy from ATP to *activate* molecules so that they can undergo a further step of chemistry. This means converting them to a derivative form that possesses adequate reactivity while also being recognizable (able to be bound) by enzymes responsible for subsequent steps.

ATP Can Serve as Both Activator and Handle

Aminoacyl transfer RNA synthetases (aa-tRNA synthetases) prepare amino acids for incorporation into proteins by esterifying the amino acid's carboxyl group with a hydroxyl group of the ribose moiety at the 3'-end of transfer RNA (Figure 8-1).

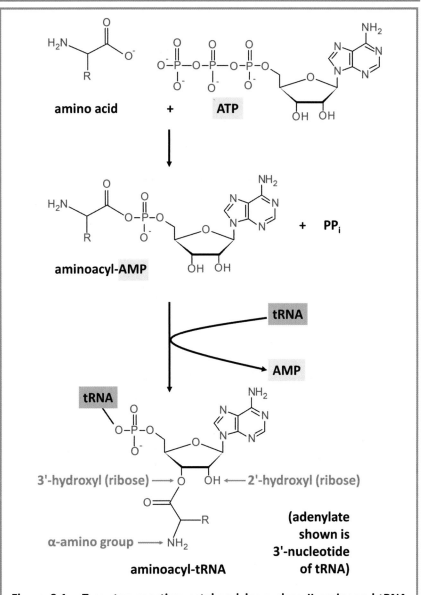

Figure 8-1. Two-step reaction catalyzed by a class II aminoacyl-tRNA synthetase. The carboxyl group of the amino acid requires activation by ATP (first step) in order to be able to form the aminoacyl-tRNA (second step). PP_i is pyrophosphate. Class I enzymes initially place the amino acid on the 2'-hydroxyl of the ribose moiety of the terminal nucleotide of tRNA, and it is shifted to the 3'-hydroxyl by transesterification.

Each aa-tRNA synthetase catalyzes a two-step reaction.

In the first step, ATP and the amino acid are condensed to form an aminoacyl adenylate (aminoacyl-AMP in Figure 8-1: an acylphosphate) with the elimination of pyrophosphate (PP$_i$).

In the second step, class II synthetases esterify the carboxyl group of the aminoacyl moiety with the 3'-hydroxyl group of ribose in the 3'-adenylate nucleotide of tRNA, yielding an aminoacyl-tRNA (Figure 8-1). Class I synthetases couple it to the 2'-hydroxyl group, but transesterification then shifts it to the 3'-hydroxyl group.

Hydrolysis of one molecule of ATP to AMP and pyrophosphate during the first step makes the overall process of aminoacyl-tRNA synthesis thermodynamically favored. This is a typical unit of chemical work performed by the cell using an ATP molecule.

In addition to the activation aspect, the ATP-derived adenylate linked to each amino acid by its specific tRNA synthetase constitutes a molecular handle or adapter for the aminoacyl moiety, which can be small. Each amino acid has its own tRNA synthetase (some have more than one), and the enzyme family is more diverse than might be expected for a class in which all members perform closely related functions.

Continuing to Protein Synthesis

A particular aminoacyl-tRNA is required for each step of peptide chain extension at the ribosome, where the C-terminus of the nascent ("being born") peptide chain (Figure 8-2) is linked to tRNA by an ester bond exactly like the one synthesized in Figure 8-1.

The α-amino group of the amino-acid moiety of the incoming aminoacyl-tRNA attacks that ester bond to make a new peptide bond, making the chain one amino-acid residue longer.

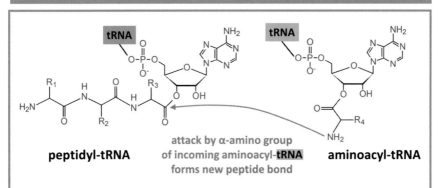

peptidyl-tRNA   attack by α-amino group
of incoming aminoacyl-tRNA   **aminoacyl-tRNA**
forms new peptide bond

Figure 8-2. The importance of chemical activation in protein synthesis. The growing peptide chain (left, shown as three amino-acid residues long) terminates in an ester bond to the 3'-end of tRNA. The α-amino group of the incoming aminoacyl-tRNA attacks this ester to form a peptide bond, extending the chain to four residues as it is transferred to the incoming tRNA.

Chemical activation of cellular components is a necessity, but it comes with the risk that they will undergo unwanted reactions. For example, the ester bonds of aminoacyl-tRNA's are at risk of hydrolysis and require protection by protein factors like the abundant prokaryote elongation factor Ef-Tu. Interestingly, its function requires not ATP but the related nucleoside triphosphate GTP (guanosine triphosphate), which Ef-Tu binds and hydrolyzes in the course of its functional cycle. Ef-Tu delivers aminoacyl-tRNA to the ribosome A-site, after which its α-amino group attacks the ester bond connecting the growing peptide chain to the tRNA in the P-site (Figure 8-2), forming a new peptide bond to extend the chain.

Coenzyme A: A Versatile Adapter

The roles of some molecules that populate biochemistry are easy enough to understand, but others appear to be there just to make the subject complicated. Coenzyme A (Figure 8-3) is an example. Accounts of metabolism have it running on and off the stage, dashing into the action only to be written out of the plot before the final scene.

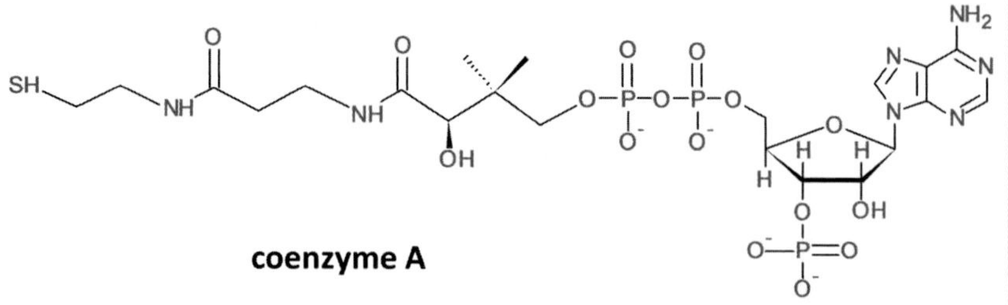

Figure 8-3. Structure of coenzyme A (CoA-SH). Its thioesters with intracellular carboxylic acids are reactive intermediates that facilitate coupling the acids to alcohols, including glycerol. CoA-SH is also an essential cofactor for fatty acid synthesis and the oxidation of pyruvate in the tricarboxylic acid cycle.

As a cofactor for many enzyme-catalyzed processes, coenzyme A does something very important by enabling the cell to use its resources of energy in a controlled, stepwise fashion. It also serves as a molecular handle or adapter that allows enzymes managing fatty acids and other carboxylic metabolites to deal alike with different but related substrates.

What Happens to Excess Energy

Building and organizing the cell takes work. That energy mainly comes from ATP, but also from the reduced coenzymes NADH and NADPH, which hold electrons at high potential and release them to perform essential reductions. These essential power sources are themselves charged by metabolic capture of energy from nutrients or, in photosynthetic organisms, light.

When the cell has more energy on hand than it needs, it stores it as fat. Evolution has naturally favored the survival of organisms that can do this, even if some modern humans find it unhelpful. Triacylglycerols (triglycerides) are a major form of stored fat. They are assembled by condensing one molecule of glycerol with three molecules of fatty acid by way of three ester bonds (Figure 8-4).

Fatty acids are carboxylic acids, and glycerol is an alcohol with three hydroxyl groups. Water is lost when a carboxyl group and a hydroxyl group combine to form an ester.

$$R_1\text{-COOH} + \text{HO-glycerol} = R_1\text{-CO-O-glycerol} + H_2O$$

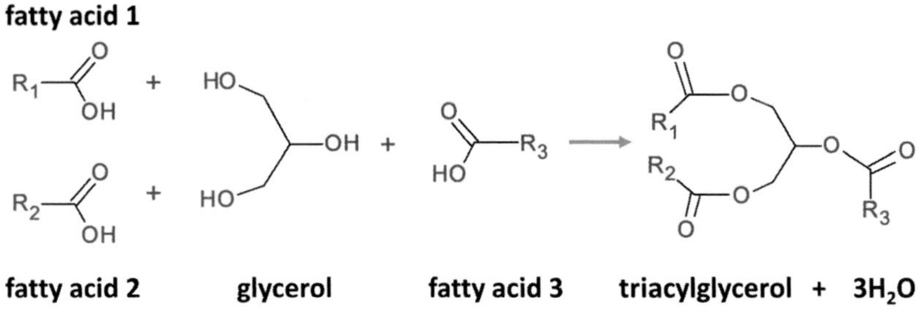

fatty acid 1

| fatty acid 2 | glycerol | fatty acid 3 | triacylglycerol + 3H₂O |

Figure 8-4. Precursors (left) and product (right) in synthesis of a triacylglycerol. The fatty acid precursors require ATP-consuming enzyme-catalyzed activation to their CoA thioesters for this synthesis to occur.

Mixing fatty acids and glycerol together under mild conditions produces no reaction. As we will see below, it takes an investment of energy, several steps of enzymatic catalysis and the participation of CoA-SH as an adapter to effect triacylglycerol synthesis.

Synthesis of Fatty Acids

Before it can make triacylglycerols, the cell first requires fatty acids. In eukaryotes, fatty acids are produced by the action of two cytoplasmic multifunctional enzymes, acetyl-CoA carboxylase (ACC) and fatty acid synthase (FAS). The carbon atoms required to build the fatty acids come from acetyl-CoA, which is abundantly produced from glucose by way of pyruvate in a well-nourished cell. Each enzyme brandishes a swinging arm derived from a B vitamin, biotin in ACC and phosphopantetheine in FAS. These prosthetic add-ons either endow the enzyme with a chemical functionality that cannot be supplied by any amino-acid side chain (biotin in ACC) or allow it to move substrates between adjacent active sites while keeping them covalently bound (phosphopantetheine in FAS).

Reducing power from NADPH is required in the course of fatty acid synthesis. A major source of NADPH is the pentose phosphate pathway, in which two distinct steps yield the reduced coenzyme.

Wisdom of the Squirrel

We can speak of energy being "invested" in fatty acid synthesis, but this is not the sort of investment that pays out more than was initially put in. The cell has evolved to spend energy freely when it is plentiful *so that reserves are available when it is not.* This invisible (to our eyes) biology has a parallel in the activities of animals like squirrels, which forage to store food reserves in the fall so that they have something to eat in the winter. The energy recovered by metabolizing stored fat cannot equal the energy required to make it, but the important thing is that it is there when it is needed.

From Fatty Acids to Triacylglycerols

Back to chemical activation. To make this synthesis possible, one of the reagents – it happens to be the fatty acid – must be given more than its own intrinsic reactivity. Again, energy stored in ATP is spent to achieve this, and the process has two steps. Coenzyme A takes the stage in the second.

First, a fatty acyl CoA synthetase catalyzes a reaction between fatty acid and ATP in which pyrophosphate is displaced from ATP and fatty acyl-AMP is formed.

$$R_1\text{-COOH} + \text{ATP} \quad \rightarrow \quad R_1\text{-CO-AMP} + \text{PP}_i$$

The acylphosphate bond between AMP and the fatty acid is reactive enough to undergo enzyme-catalyzed attack by the thiol group of CoA-SH, forming a thioester product R_1-CO-S-CoA.

$$R_1\text{-CO-AMP} + \text{HS-CoA} \rightarrow R_1\text{-CO-S-CoA} + \text{AMP}$$

Thioesters are labile compounds: they are reactive to the attack of a nucleophile at the carbonyl function of the thioester, with the thiol component being driven out and a new bond formed between the carbonyl carbon and the attacking molecule. (The term *thio* in the name of a compound always means that sulfur has replaced oxygen in the structure denoted by the remainder of the name).

Enzymatic catalysis steers the fatty acyl-S-CoA to its end-use. In triacylglycerol synthesis, the final step in assembly is catalyzed by either of two forms of an enzyme called diglyceride O-acyltransferase (DGAT: the two forms are DGAT1 and DGAT2):

diacylglycerol + fatty acyl-S-CoA → triacylglycerol + HS-CoA

Again, making a labile thioester from the fatty acid and CoA-SH is a way of maintaining the fatty acid in a sufficiently reactive form to allow it to be linked to the free hydroxyl group of diacylglycerol.

Maintaining a degree of chemical activation – keeping an otherwise unreactive group ready to be linked to a group that attacks it – is the first and principal function of coenzyme A. ATP provides the energy, but coenzyme A is the adapter that enables the necessary metabolic operations. Carboxyl groups are the beneficiaries of this essential contribution to metabolism.

Coenzyme A is quite a large group, but it has the important function of being linked to the small (two-carbon) acetyl unit in the key metabolite acetyl-CoA. Evolution may have selected this derivative as one that allows the two-carbon unit to be recognized and managed more effectively than would be the case if its reactivity came from thioesterification with a smaller thiol.

Activation to Assist Bond-Making in the Laboratory

Finally, it is often useful to biotinylate a protein reagent to allow it to be immobilized on streptavidin-coated beads or to make it readily detectable with a streptavidin-linked reporter.

Also, as we saw in Chapter 7, covalent attachment of fluorescent groups to proteins or other biomolecules is frequently required.

The snag is that biotin (for example) is a carboxylic acid, and suffers from the same low reactivity under mild conditions as other compounds in that class. That can be remedied by converting biotin to a succinimidyl ester, an activated form that reacts under mild conditions with protein amino groups. Biotin may have an extra spacer arm added, as shown in Figure 8-5.

As we should now be used to, N-hydroxysuccinimide is relatively easily bumped out of the carboxyl group by a nucleophilic amino group that replaces it. The result is formation of a stable amide bond (Figure 8-5). Other nucleophilic groups present in proteins may also attack the reagent, but form labile adducts that undergo hydrolysis.

In biochemistry, we are very often dealing with reactions that need to take place in the presence of water. Hydrolysis of the key reagent can compete with the process that we want.

In this circumstance, chemical activation of a carboxyl group requires a nice balance of properties. The activating modifier needs to be stable enough to resist rapid hydrolysis, but reactive enough to allow for an adequately quick reaction when it is attacked by the desired final partner. Both in the cell and in the lab, this principle finds many applications.

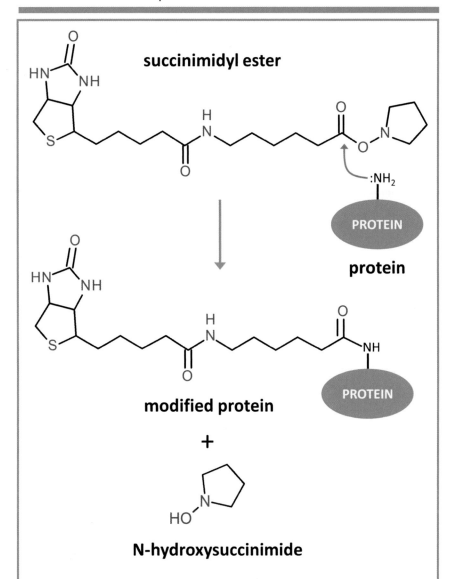

succinimidyl ester

protein

modified protein

+

N-hydroxysuccinimide

Figure 8-5. Use of a biotin succinimidyl ester (with extra spacer) to modify an amino group on a protein. Commercially available reagents of this kind are widely used for protein derivatization (see Figure 7-2). Their moderate reactivity allows N-hydroxysuccinimide to be readily displaced by an attacking nucleophile such as a protein amino group at pH~8 without the reagent being excessively consumed by the competing reaction of hydrolysis. An ethyl ester, in contrast, would lack the requisite reactivity.

Redox – What's It All About?

9

Electrons in Motion, and Redox Reactions at Metals

At a simple level, we understand electricity. Electrons are charged particles that can be lifted to and stored at a high potential energy, as in a battery. Later, they can be released to run to a state of lower potential while the freed energy is exploited for work or heat.

Redox (oxidation and reduction) reactions of metal ions may also be familiar from inorganic chemistry. For example, if Cu^{2+} is converted to Cu^{3+}, the copper(II) ion has been oxidized. The reaction increased its oxidation number, the number of electrons it has lost relative to the free element, and the electron that Cu^{2+} gave up went to a reaction partner that we call an oxidant. The oxidant was thereby reduced: its oxidation number was decreased. (Recall also that the oxidation number can be negative if the atom has gained one or more electrons relative to its free state).

Some important biochemical redox reactions are like this. The energy-harvesting electron-transport chain in the mitochondrion includes cytochromes, proteins that contain iron bound to an organic heme group. Cytochromes go through a redox cycle in which they sequentially acquire an electron (undergo reduction) from a partner upstream in the chain, then transfer it (undergoing oxidation) to the next molecule in the chain, which is reduced.

Energy extracted from this "downhill" flow of electrons from high to low potential is converted into what is called a transmembrane gradient of proton concentration (actually pH) that drives the synthesis of ATP mediated by ATP synthase.

Organic Molecules

Things are less clearcut when we come to redox reactions taking place on small or large organic biomolecules, but these are immensely important and we need to understand them.

The living cell constantly refines energy into usable forms. ATP is the main one, as we have said more than once, but the reduced coenzymes NADH and NADPH are also important. Enzymes couple the discharge of potential energy stored in these molecules to driving reactions required for metabolism. Whenever electrons are transferred from one molecule to another, the molecule that gains electrons is said to be reduced and the molecule that gives them up is said to be oxidized.

In this chemical terminology, electrons do not migrate between reactants as naked subatomic particles. Draw out the reaction, and you will often see hydrogen atoms being added to or subtracted from the substrate.

As to whether the substrate has been reduced or oxidized, the key is whether the atom at the reaction center finishes with more or fewer electrons clustered around it. The difference either way is often two, the number required to make a single bond.

To anyone hoping for simple, self-explanatory descriptions, the terminology around reduction and oxidation can be challenging.

A second definition of oxidation is the addition of oxygen to a molecule: oxygen is very electronegative, and draws electrons toward itself at the expense of the atom at the reaction center. In this sense, a reaction that places oxygen next to a reference atom deprives the reference atom of electrons, and a reaction that removes oxygen restores them.

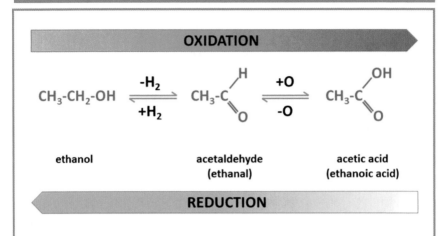

Figure 9-1. Redox chemistry involved in converting an alcohol to a carboxylic acid and vice versa. Note how the loss of H_2 and the addition of an oxygen atom are both called oxidations. In the reverse conversion, the removal of oxygen and addition of H_2 are both termed reductions.

Redox States of the Hydroxyl and Thiol Groups

Some helpful Illustrations come from the redox chemistries of carbon alcohols, R–OH, and thiols, R–SH.

Oxidizing an alcohol to an aldehyde requires loss of H_2 from the alcohol (Figure 9-1). In the forward direction of the reaction catalyzed by alcohol dehydrogenases (alcohol to aldehyde), H_2 from the substrate is transferred to the coenzyme NAD^+ and solvent H_2O, giving NADH and H_3O^+.

Further oxidation of acetaldehyde to acetic acid by aldehyde dehydrogenases similarly involves the transfer of H_2 to $NADP^+$ and solvent, but in this case the hydrogen comes from water rather than the substrate. Water is also the source of the oxygen added into the substrate. (Some aldehydes are oxidized to acids by aldehyde oxidases, a different type of enzyme that uses molecular oxygen as oxidant and generates hydrogen peroxide as a product).

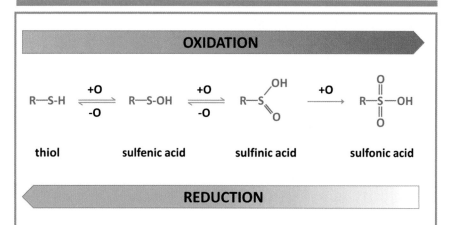

Figure 9-2. The thiol group and products of its stepwise oxidation. The redox chemistry of thiols is often nonenzymatic. Each step in either direction requires another reactant, either an oxidant (left to right) or a reductant (right to left). Conversion to the sulfonic acid is shown as irreversible.

The thiol group –SH is exceptionally important in biochemistry, and its redox chemistry (Figure 9-2) has a lot to do with its prominence. It exists both in proteins, where it is the most chemically reactive group at neutral pH, and in small molecules. An important example of the latter is glutathione, a cytosolic 305 Da thiol compound that contributes importantly to maintaining the proper redox environment within the cell and to protecting the cell against damaging oxidants. It exists in cells at concentrations on the order of 10 mM as a mixture of its thiol and disulfide (S–S) forms.

Whenever you consider a redox reaction, make it a habit to be aware of both ends of the transaction. If one reactant is reduced, another must be oxidized – the electrons transferred into the substrate being reduced must come from somewhere.

In this regard, it makes no difference whether the reaction is enzyme-catalyzed or not. Both kinds of redox process are important in biochemistry.

A Redox Reaction in Metabolism

As a specific example, let's take the reaction catalyzed by glucose 6-phosphate dehydrogenase (Figure 9-3).

The coenzyme, in this case $NADP^+$, and solvent receive the equivalent of H_2. The name of the enzyme – a dehydrogenase – tells us that it catalyzes a redox reaction.

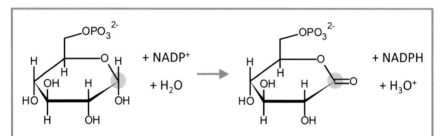

Figure 9-3. The redox reaction catalyzed by glucose 6-phosphate dehydrogenase. Carbon-1 of glucose 6-phosphate (yellow highlight) loses H_2 to the coenzyme $NADP^+$ and the solvent.

Do not assume that the enzyme's name always indicates the biologically significant direction of the reaction (although for glucose 6-phosphate dehydrogenase, it does). An enzyme has to be able to catalyze the same reaction by the same mechanism *in both directions*, and an enzyme named as a dehydrogenase may be more important in catalyzing a reduction for which coenzyme and solvent donate H_2.

Disulfide Bonds, Secreted Proteins and Intracellular Proteins

Proteins that are secreted from the cell – into the blood, for example – very often contain disulfide bonds. These bonds link the respective thiol sulfur atoms of different cysteine residues in the protein, and tend to stabilize a protein.

In contrast, the environment within the cytoplasm is generally described as reducing relative to the status of thiol groups, and cytoplasmic proteins mostly contain free thiol groups rather than disulfide bonds.

Let's examine the oxidation of thiols and reduction of disulfides more closely. A single covalent bond exists when two electrons occupy a molecular orbital between two atoms. In the thiol group R–SH, we focus on the bond between sulfur and hydrogen. In each thiol, it is formed by the two electrons represented by dots in Figure 9-4.

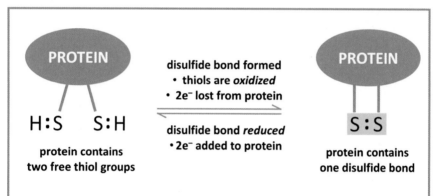

Figure 9-4. "Reduced" and "oxidized" states of a protein in which two cysteine residues can form a disulfide bond. Formation of the disulfide (yellow highlight) requires that two electrons (2e⁻) are transferred from the protein to an oxidant, which by this process is reduced. Reduction of the disulfide requires that two electrons are donated to the protein.

After oxidation, a single bond between the two sulfur atoms has taken the place of the two sulfur-to-hydrogen bonds that were present in the two thiols. Two electrons have been stripped out of the molecule and taken away by the oxidant.

A good oxidant is a compound that is easy to reduce. One example is hydrogen peroxide (Figure 9-5).

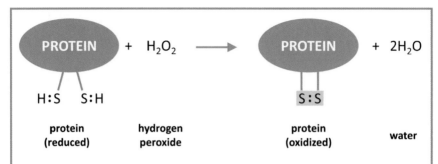

protein (reduced) hydrogen peroxide protein (oxidized) water

Figure 9-5. Oxidation by H_2O_2 of a pair of thiol groups in a protein to form a disulfide bond. The thiol groups have been oxidized – a net transfer of two hydrogen atoms has occurred to hydrogen peroxide, the oxidant, giving two molecules of water. The oxygens in H_2O_2 have been reduced.

Next, let's examine reduction of a protein disulfide bond by dithiothreitol (DTT), a reagent in which the reduced form contains two thiol groups.

As shown in Figure 9-6, the protein and the reducing agent exchange states in a way that is easily seen. DTT initially has two free thiols, and finishes with an intramolecular disulfide. The protein undergoes the inverse of this switch.

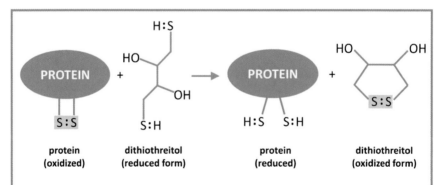

protein (oxidized) dithiothreitol (reduced form) protein (reduced) dithiothreitol (oxidized form)

Figure 9-6. Reduction by DTT of a disulfide bond in a protein to produce two free thiol groups. A net transfer of two hydrogen atoms has occurred to the protein, and dithiothreitol is left with an intramolecular disulfide bond.

Why should this reaction work? Why would DTT undergo oxidation while the protein gets reduced, with the two molecules exchanging their respective conditions?

By the same logic that makes H_2O_2 a good oxidizing agent, *a good reducing agent is a compound that is easily oxidized.* Because DTT undergoes an *intramolecular* oxidation to form a 6-membered ring, its oxidation is a favored reaction. Yes, the protein disulfide is also intramolecular in a formal sense, but the creation of a 6-membered ring is an especially favored process in chemistry. Therefore, DTT easily undergoes conversion from its reduced state to its oxidized state while the protein undergoes the reverse.

Analytical protein chemists use this reaction (accelerated by heat) to force the reduction of disulfides. DTT is also used as an additive to protein solutions during purification or storage, when it is intended to act as a scavenger for oxidants that might otherwise cause chemical change in the protein.

As noted above (Figure 9-2), disulfide formation is not the only type of oxidation that thiols undergo in proteins. R–SH can be oxidized progressively to three different oxidized states (Figure 9-7).

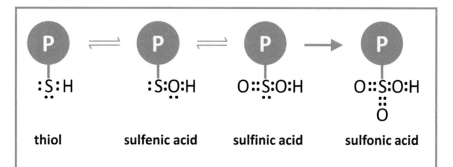

Figure 9-7. Three oxidized derivatives of the thiol group. The names of these states can be remembered using a mnemonic, as they differ by a single letter that changes in alphabetical order from "e" to "i" to "o."

Considering the different oxidation states shown in Figure 9-7, we see that the oxidations beyond the level of the thiol are not simply a matter of electrons being taken away, but rather due to an additional oxygen atom entering the molecule at each step.

Everyday life tells us quite a bit about oxidation and reduction. The combustion of wood in air is oxidative with respect to the wood and reductive with respect to oxygen as the carbohydrates in the fuel combine with atmospheric O_2 to form CO_2 and H_2O. Energy is liberated – we can warm ourselves at a camp fire – so we know that the reaction is exothermic, and the reaction's strong tendency to continue once started means that it is thermodynamically favored.

Our own tissues conduct the same oxidation of food carbohydrates, but in a controlled, stepwise manner so that the energy extracted is captured for later use. Instead of being combined directly with oxygen and releasing all of the stored energy in a single spectacular step, nutrient carbohydrates are carefully disassembled in stages so that the energy that they embody can be converted into biologically useful forms.

In contrast, the reverse reaction of combining CO_2 and water to make glucose while liberating O_2 – the process of photosynthesis – consumes energy and is only possible because plants have evolved to harvest energy from sunlight.

Marvelous Mass Spectrometers

10

Technology is by definition not magical, but it can be so obscured by the jargon of its initiates that it might as well be based in wizardry. A prime example of this is mass spectrometry.

A range of devices and strategies exist that are variously selected to provide optimal analyses and quantitation of all types of biomolecule. Here we focus on work with proteins and peptides. The need to name these systems and methods succinctly cooks up an alphabet soup of acronyms that can intimidate the newcomer. Some complexity is unavoidable, but appreciating the common features of all mass spectrometers is a great way to manage it.

Mass spectrometers must solve three problems (Figure 10-1A).

1. **Get the analyte into the gas phase, usually as a mixed population of positively-charged ions**

2. **Resolve ions of different mass-to-charge ratio (m/z)**

3. **Detect ions while registering their m/z values**

Each challenge can be met in different ways. Engineers designing a mass spectrometer select one device to handle each distinct phase of the process, then put them together in a chain.

Mass spectrometry started with F. W. Aston (Cambridge, U.K.) early in the twentieth century, but was not useful with fragile biomolecules until methods were discovered to guide them into the gas phase without tearing them apart. Two good solutions to this problem emerged in the nineteen-eighties, earning shares of a Nobel Prize in Chemistry for J. B. Fenn and K. Tanaka in 2002.

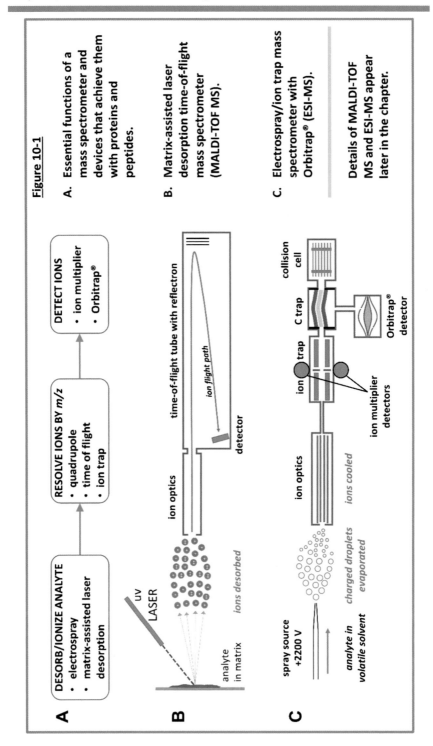

Figure 10-1

A. Essential functions of a mass spectrometer and devices that achieve them with proteins and peptides.

B. Matrix-assisted laser desorption time-of-flight mass spectrometer (MALDI-TOF MS).

C. Electrospray/ion trap mass spectrometer with Orbitrap® (ESI-MS).

Details of MALDI-TOF MS and ESI-MS appear later in the chapter.

Desorption and Ionization

As we saw in Chapter 5, water-soluble proteins and peptides interact intimately with their aqueous environment. Acidic and basic groups on the molecule are variably ionized according to the pH. For example, at neutral pH the carboxylic side chains of aspartyl and glutamyl residues exist as $-COO^-$ while amino groups of lysyl residues are protonated and positively charged as $-NH_3^+$.

Radical change occurs when a molecule is stripped of solvent and delivered into the gas phase as singly- or multiply-protonated and charged positive ions. Except for the sites at which protonation is present – it is needed to create charge – groups that are usually ionized in solution adopt their neutral form when desorbed from the solvent (Figure 10-2). Carboxyl groups become protonated and neutral as $-COOH$, while amino groups and other sites that are protonated in solution can lose their protonation in the gas phase and become neutral.

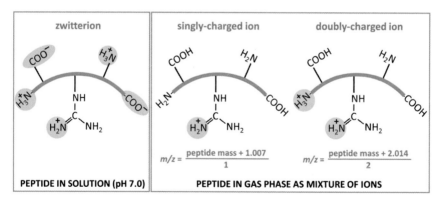

Figure 10-2. The effect of migration from aqueous solution (left) to the gas phase (right) on typical protonation states in a peptide. Depicted protonation states should not be taken as fixed or exclusive of others. Key learning: a pure peptide frequently enters the gas phase *as a mixture of differently-protonated and differently-charged ions*. Therefore, the mass spectrum of a single, pure peptide may contain multiple peaks, as shown in Figure 10-3A.

Mass Spectrometers Measure Mass-to-Charge Ratio (*m/z*)

One key to understanding scientific instruments is to know what they actually measure. For example, we use a clinical thermometer to measure body temperature, but the relative expansion and contraction of the alcohol in its bulb is the physical quantity that we observe. Likewise, we use mass spectrometers to measure molecular mass, but the quantity actually gauged is the mass-to-charge ratio (*m/z*) of the analytes.

Mass spectrometers manage and detect ions by subjecting them under high vacuum to rapidly fluctuating electric fields and the electrostatic influences of electrodes. Unit electric charge (or its multiples) will experience a fixed level of attraction or repulsion to or from any electrode, but the mass of the molecule carrying it will determine how it moves in response. Molecules with different masses could, with suitably different protonation, present identical *m/z* values. Fortunately, nature and science combine to allow us to surpass this problem and confidently extract masses from peaks registered on a scale of *m/z* (see below).

Derivations of *m/z* for singly- and doubly-charged peptide positive ions appear in Figure 10-2. For gas-phase peptide molecules that would otherwise be neutral, charge comes from protonation. The quantity 1.007 Da is a rounded value for the mass of H^+ (1.007276 Da). For a twice-protonated doubly-charged molecule, $[M+2H]^{2+}$, the mass contribution of 2H is twice this value, or 2.014 Da.

Mass-to-Charge Ratio (*m/z*).......Does It Have a Unit?

Although it seems illogical, *m/z* is usually reported with no unit as "dimensionless." After all, the unit of molecular mass is the dalton (Da), equivalent to 1.6605×10^{-27} kg, and the elementary electric charge *e* counted in the denominator of *m/z* equals 1.602×10^{-19} C (coulombs; the coulomb is equivalent to one ampere-second). A unit called the thomson (Th) has been proposed to denote m/z systematically (Da e^{-1}), but has yet to be widely adopted.

Isotopic Diversity Allows Us to Deduce the Value of z in m/z

98.9% of the carbon atoms in your body are ^{12}C, the most abundant isotopic form, but 1.1% are the stable isotope ^{13}C. In a peptide containing around 80 carbon atoms, there is a high probability that at least one carbon somewhere in any molecule will be ^{13}C, so that the chemically pure peptide will be a mixture of some molecules in which all the carbons are ^{12}C and others that contain one, two or more atoms of ^{13}C. Each form has a different mass (Figure 10-3B).

They differ in mass by increments of 1 Da. The lightest molecules contain only ^{12}C, and their mass is called the *monoisotopic* mass. Close viewing of the peak cluster with monoisotopic m/z of 800.938 (Figure 10-3B) shows that the peaks are separated by 0.5, telling us that z must be 2 and allowing us to extract the monoisotopic mass (as in Figure 10-2) even if this peak cluster is the only one detected.

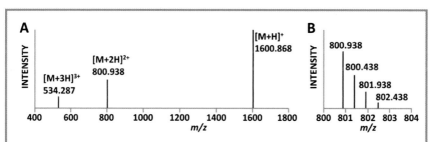

Figure 10-3. Mass spectrum of a 1599.862 Da peptide and close-up of a multiply-charged peak within it. **(A)** The pure peptide gives more than one peak because the desorption/ionization process generates differently protonated/charged forms. **(B)** A close-up view of the peak at $m/z=800.938$ (monoisotopic) shows that it is actually a cluster of peaks representing different isotopic forms of the peptide. These have m/z separated by 0.5 while their respective masses are separated by 1 Da, so z for this peak group must be 2. Peaks in the cluster for [M+3H]$^{3+}$ are separated by 0.333 units.

MALDI-TOF MS

In *matrix-assisted laser desorption-ionization* (MALDI) (Figure 10-1B), analyte is mixed with a saturated solution of a matrix compound, which is a uv-absorbing organic acid such as α-cyano-4-hydroxycinnamic acid. A drop of the mixture is allowed to dry, and crystals of the matrix form with the analyte entrapped.

Under vacuum in the front compartment of a typical instrument, the dried sample receives pinpoint pulses of irradiation from a near-uv laser. The acidic matrix readily absorbs light at the laser wavelength (e.g. 337 nm), and the tiny irradiated region of the sample undergoes extreme energization that causes sublimation of matrix molecules (transfer to the gas phase). Along with this are effects on the entrapped analyte biomolecules, some of which are delivered intact into the gas phase as protonated positively-charged molecules. These ions are the usual subjects for mass spectrometry, although we may occasionally be interested in negative ions or in fragment ions created from the analyte.

MALDI has a useful tendency to produce singly-protonated singly-charged ions. These are designated $[M+H]^+$, indicating that they correspond to the neutral molecule M carrying one proton H^+ and a single overall positive charge. Admitting these to the detector allows their respective mass-to-charge ratios to be determined.

A time-of-flight (TOF) analyzer/detector is often used with MALDI ionization (Figure 10-1B) to accomplish the second and third critical tasks noted in Figure 10-1A. After each laser shot, positive ions supplied to the TOF device are focused by "ion optics" into a tight packet and accelerated into the high-vacuum flight tube. Consider a set of singly-charged positive ions with a range of masses, which we will reduce to just two, one X^+ that is lighter than the other XX^+.

Each receives an equal accelerating force, but the lesser mass (lower inertia) of X^+ causes it to fly faster down the tube than XX^+. If we record the time of flight for each ion and compare these times with flight times of calibrants of known m/z, we can plot ion intensity vs. m/z. This is our mass spectrum (Figure 10-3).

Most TOF analyzer/detectors contain an ion-focusing device called a reflectron (Figure 10-1B) that improves their resolution. As discussed in Chapter 12, resolution relates to peak width. Narrow peaks like those in Figure 10-3B allow the instrument to distinguish m/z values that are very close to each other.

Key learning: MALDI-TOF MS is a process of "irradiate, collect and repeat." Hundreds, sometimes thousands, of repeats of the pulse-and-detect routine are typically needed to collect a spectrum, but modern MALDI-TOF instruments manage this in seconds or less.

MALDI's propensity to create singly-charged ions makes it a good method for analyzing mixtures of peptides, such as tryptic digests of pure proteins or families of related peptides such as β-amyloid.

Finally, here is a real-world analogy to an aspect of MALDI-TOF MS. You and your identical twin go to the bowling alley. You select a 10 lb bowling ball, but your twin (whose physique and condition match your own) prefers a 16 lb ball. You each apply identical propulsive force to your chosen ball. The 10 lb ball travels more quickly to the pins than the 16 lb ball, because the same force applied to less mass results in greater acceleration (acceleration = force/mass).

In TOF MS, when positive ions are accelerated down the flight tube by electrostatic lensing, ions with less mass per unit charge (m/z) complete the trip more quickly than those with greater m/z. Calibration is achieved by recording the flight times of standards.

ESI MS

Electrospray ionization (ESI) (Figure 10-1C), the dominant front-end strategy for protein/peptide MS, is invariably used in proteomics. It can be coupled to a range of different mass detectors including TOF, ion trap and Orbitrap® devices.

For ESI to succeed, analytes must be freed from nonvolatile components. This can be accomplished using reversed-phase liquid chromatography, which delivers peptides or proteins in water mixed with a volatile organic solvent (often acetonitrile, CH_3CN) and a dilute volatile acid (often 0.02-0.1% formic acid). Pumping this solution through an electrified narrow-bore needle converts it to a fine spray of positively charged droplets. Evaporation of neutral solvent molecules from the droplets leads to the surface tension holding them together being overpowered by the mutually repulsive positive charges on their surface, causing them to disintegrate. Protonated, positively charged analyte molecules enter the gas phase, and are attracted by charge difference into the mass-measuring module of the spectrometer.

In contrast to MALDI, ESI tends to form multiply-charged ions from proteins and peptides, $[M+n(H)]^{n+}$, where n is the number of protons attached to the otherwise neutral molecule (Figure 10-2).

This leads to a situation of "one protein, many peaks", so that the spectrum of a single component often looks more complex than the one produced after MALDI.

A pure, single protein gives a spectrum with many peaks, but the calculation of protein mass from its spectrum can be done by hand or with a calculator. It usually makes better use of the data to leave this to a computer, but you should understand the process.

Suppose that our mass spectrometer operates with a *m/z* range of 400-2000, which is not unusual. When we analyze a 20,000 Da protein, ESI has to generate multiply-charged ions with sufficient charge to bring their *m/z* into the instrument's range. For example, with 20 protons attached to the protein, the ion $[M+20H]^{20+}$ will have *m/z* of 1001.0 and our instrument can detect it.

The same process inevitably produces ions with similar counts of protonation and charge: along with $[M+20H]^{20+}$ will come ions on either side of that value (Figure 10-4). Be sure to understand that *increasing* protonation/charge states result in *decreasing* values of *m/z*: the charge state *z* is the denominator of *m/z*.

Now we seem to have a problem. Knowing the value of *z* for even one peak should allow us to calculate the protein mass – but how are we supposed to know this value? Nothing in the appearance of the spectrum gives it away.

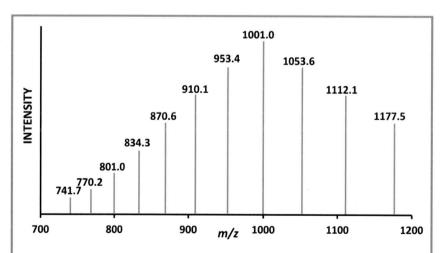

Figure 10-4. Theoretical ESI mass spectrum of a 20,000 Da protein. Note that (i) peaks in the spectrum are *not* equally spaced, and gaps between them become narrower as *m/z* falls; (ii) each peak provides a separate estimate of the protein mass, but it is normal to average out the separate measurements to obtain the final result. The text shows how the mass can be extracted.

As mentioned above, the problem is simpler than it looks and can be solved using a calculator or with pencil and paper. *The key is realizing that the peaks in the spectrum are members of a geometric series.*

If one peak in the spectrum corresponds to an unknown protonation/charge state n, the next one to the left (lower *m/z*) must have the protonation/charge state (n+1).

Then the respective *m/z* values for those peaks must equal (M+n)/n and (M+n+1)/(n+1).

Algebra allows us to solve these simultaneous equations for the value of n, and this enables us to calculate the protein mass M (Figure 10-5).

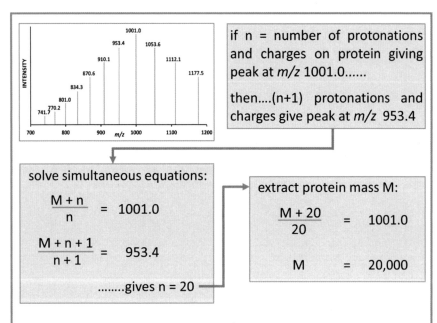

Figure 10-5. **Each multiply protonated/multiply charged peak in a protein mass spectrum provides an estimate of the protein mass. For an optimum result, information from all peaks is combined. This process is usually computer-based.**

Detectors

Once ions have been produced from the analyte, the mass spectrometer has to fractionate them according to their various *m/z* values and then register their presence. The classical workhorse detector based on an electron multiplier has the sole function of registering the arrival of ions, with the task of resolving ions according to *m/z* values handled upstream by a device such as the TOF flight tube.

Another mass-resolving device is the three-dimensional ion trap, which "cools" or energizes ions by radio-frequency variations of an electric field created in a quadrupole. Ion traps typically have electron multiplier detectors associated with them, although in modern multistage instruments, their use is optional. Ions may be forwarded from the trap to a higher-performance detector.

With regard to resolution – the ability to distinguish ions with *m/z* values very close together – the most powerful detector remains the Fourier transform ion cyclotron. However, these detectors include a large magnet that requires careful management, and their slow duty cycle is not ideal for proteomics.

A very important and successful alternative, commercialized in 2005, has been the Orbitrap® detector. Along with high-performance TOF systems, this compact device has allowed rapid progress in proteomics and other aspects of biochemical mass spectrometry.

Web sites and literature from instrument vendors provide excellent information about these systems, often including animations.

Molecular Interactions by Surface Plasmon Resonance

11

Specific interactions between molecules underpin the chemical basis of life. Once Emil Fischer had proposed the lock-and-key hypothesis for enzyme-substrate interactions in 1894, visualizing, refining and generalizing his insight was a dominant goal of twentieth-century biochemistry.

Biochemists worked for decades to understand how molecules bound to each other by determining how they looked either alone or in a complex. With x-ray crystallography to the fore, this structural work confirmed that immensely particular complementarity between molecules allows them to bind specifically to each other, even at low concentrations and in the presence of many other molecules. Small molecules often occupy pockets on large molecules, while large molecules usually interact with each other across larger binding surfaces.

Beyond enzymology, less attention was paid to *how quickly* molecules bind to each other or how rapidly their complexes fall apart. As in Chapter 2, we must clearly distinguish rates from rate constants. Rates refer to what actually happens, and they change according to the concentrations of the binding partners. In the limit, when one partner A is absent (has a concentration of zero), the rate of its interaction with a partner B must be zero.

A rate constant, on the other hand, describes the intrinsic "ability to happen" of a process. More formally, it expresses the relationship between the concentrations of the reactants and the rate at which the process occurs.

Detectors

Improved means to measure the kinetics of molecular interactions and dissociations have led to sharply increased awareness of their significance. In drug discovery, for example, a drug that stays bound to its target for a very long time before dissociating may have a longer duration of action than one that dissociates rapidly and is cleared more quickly from the vicinity of the target. The ability to measure these rates and compare the performances of different molecules has become highly valued.

In this regard, the concept of dynamic equilibrium is very important. The extent to which two molecules, R and L, bind to each other in a complex R-L is governed by their respective concentrations and the degree of affinity that exists between them. Their mutual affinity is gauged by their equilibrium dissociation constant, K_d:

$$K_d = \frac{[R]\,[L]}{[R\text{-}L]}$$

When we mix R and L together, the system has to reach an equilibrium at which the respective concentrations of R, L and R-L are constant. If R and L bind to each other quickly, equilibration will be fast. If they bind to each other slowly, it will take longer.

When equilibration has been achieved, the rate at which R and L are newly associating with each other has to be the same at which R-L is dissociating to give free R and free L.

We can call the second-order rate constant for association k_{on} (units: $M^{-1}\,s^{-1}$), and the first-order rate constant for dissociation k_{off} (units: s^{-1}). Then, when the system is at equilibrium:

$$k_{off}\,[R\text{-}L] = k_{on}\,[R]\,[L]$$

and

$$\frac{k_{off}}{k_{on}} = \frac{[R]\,[L]}{[R\text{-}L]}$$

so that

$$\frac{k_{off}}{k_{on}} = K_d$$

This statement has important implications for our thinking about complexes between molecules, such as those between drugs and their targets.

Consider two drugs that each bind tightly to the same target with K_d of 1 nM. k_{off}/k_{on} for each of them must equal 1 nM, but it is possible for one of them to bind to the target and dissociate from the target much more rapidly than the other, while the *ratio* of the rate constants k_{off} and k_{on} for each compound has the same value.

How will this affect their respective performances? If both compounds are equally well absorbed by the body, metabolized at the same rate, and reach the target at the same concentration, the one that binds slowly to the target should – on first dose – take longer to reach maximum activity. Importantly, though, its slower dissociation from the target will make its activity last longer.

The difference might be crucial, depending on the drug. A cure for migraine headaches needs to act quickly: nobody wants to wait eight hours for relief. On the other hand, medicines that treat chronic conditions like diabetes or depression can afford to have a slower onset of action if that action is effective once it takes hold. (Remember that the speed of a drug's onset of action is affected by many factors, such as its adsorption and metabolism. The kinetics of its binding and dissociation are only one of these).

Direct Measurement of Binding Events

In enzymology, reaction rate measurements can reveal kinetic information about events such as the binding of an inhibitor, especially for inhibitors that bind very slowly to the enzyme. For nonenzyme ligand-binding proteins, though, it was difficult for many years to obtain a real-time correlate of the binding of one molecule to another. Rapid-mixing ("stopped-flow") methods with spectroscopic readouts could sometimes fill this gap.

In 1990, the first commercial version appeared of a new generation of instruments that detect molecular interactions as they happen. Their unique centerpiece is a sophisticated measuring device called a "chip" which has two sides, wet and dry. Above a foundational stratum that includes a thin layer of gold, the wet side is coated with dextran to which one molecule of a binding pair is coupled either covalently or by a tight noncovalent association (e.g. through an antibody). A stream of liquid continually pumped over it contains either buffer or a solution of its binding partner. It is as if the immobilized molecule coats a rock that sits in an ever-flowing stream. During an experiment, the source of the stream will be switched from one reservoir to another.

Beneath the gold-coated surface, on the "dry" side of the chip, is an optical system that directs light onto the metal layer and generates within it an electromagnetic resonant wave (surface plasmon resonance, or SPR) that affects the distribution of angles at which light is deflected to a detector. This distribution changes when molecules bind to or dissociate from the layer on the wet side of the chip. The variations can be recorded in the form of a signal that is interpreted in terms of molecules binding or leaving the dextran layer. Biochemists do not need to understand this subtle process in detail, but explanations of it are widely available.

The Story of a Single Sensorgram

The multistage traces that record real-time binding and dissociation events in these instruments are called sensorgrams. At the baseline (Figure 11-1A), the liquid flowing over the chip is a buffer solution.

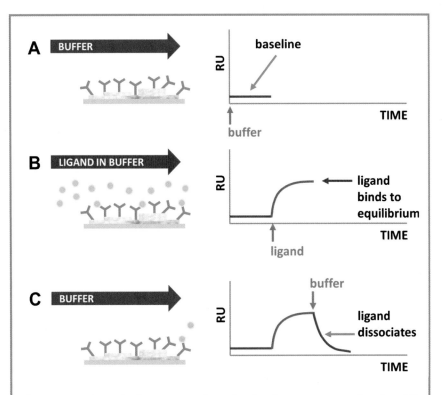

Figure 11-1. Steps in the construction of a simple sensorgram showing: (A) baseline reading as buffer flows over an immobilized protein; (B) a progressive rise in output signal as a solution containing a ligand to the immobilized protein flows over the "chip"; (C) the decline in signal when the ligand solution is replaced by buffer and bound ligand undergoes dissociation from the immobilized protein. Small red arrows indicate points at which the solution flowing over the chip is changed. The optical detection system on the dry side of the chip is not shown. It detects variations in surface plasmon resonance (SPR) in the chip, and the y-axis label "RU" denotes response or resonance units.

Next we switch the flow to a solution of the binding partner of the immobilized molecule (Figure 11-1B). As binding starts to occur, the signal from the chip changes in a way that is read as an increase, and rises until an equilibrium level of binding is achieved.

Whether or not this level corresponds to saturation of the immobilized partner depends on the effective concentrations of the two binding partners and their mutual affinity. When equilibrium is reached, binding of ligand to and dissociation of ligand from the immobilized partner are occurring at equal rates.

Remember that *liquid flows over the chip at all times*. When the ligand solution is introduced, the immobilized protein is being exposed to a constant concentration of ligand. A pre-steady-state phase is observed first in which the level of ligand bound to the immobilized protein is climbing toward the equilibrium value. The steady-state phase is a case of dynamic equilibrium.

Next, *we switch the incoming flow back to buffer* (Figure 11-1C). No new ligand binding can occur now, but dissociation of ligand into the mobile phase continues at the same rate as before. The molecular complex present on the chip dissociates, the ligand is flushed away by the flowing buffer, and the signal output from the optical system declines.

Ultimately, curve-fitting software is used to compare experimental SPR traces with models of molecular interactions and to extract both rate and equilibrium dissociation constants. Although the first SPR-based biosensors were limited to detecting interactions between macromolecules, further development has allowed the interactions of small molecules with proteins to be brought into scope. As a result, direct analyses of drug-to-target binding interactions is now a central method in drug discovery.

Variations of Ligand Concentration

Even a single sensorgram can yield information, but variations of ligand concentration are normally used to improve the accuracy of the values extracted. Progressively increasing the concentration of the free ligand should raise the height of the plateau reached at the top of the binding curve until binding is maximized (Figure 11-2). This provides one route to extracting the equilibrium dissociation constant of the interaction under study.

The up-and-down trace shown in sensorgrams from these instruments can be hard for newcomers to understand. The first phase shows a binding event; then, after a change in the solution running past the chip, the second phase shows dissociation. The two phases are usually put together as a single trace, and understanding the stepwise nature of the experiment is the only way to make "sense" of what is shown.

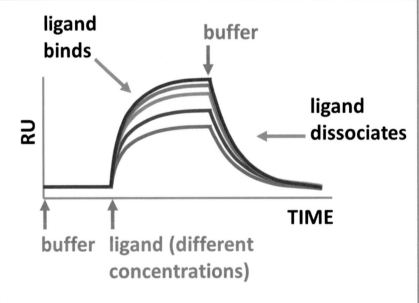

Figure 11-2. Superimposition of five sensorgrams prepared using the same immobilized protein with five concentrations of the soluble binding partner.

Conventional Chromatography

12

"He that breaks a thing to find out what it is has left the path of wisdom" – Gandalf the Grey

Gandalf was a wise and mighty wizard in J. R. R. Tolkien's tales of Middle Earth, but perhaps he lacked the instincts of a biochemist. After all, much of biochemistry rests on taking living systems apart, purifying the components, and studying them in isolation or small combinations.

An important part of this work deals with complex, delicate functional structures such as mitochondria, ribosomes and chloroplasts, the isolation of which requires specialized methods of fractionation. More foundational, though, has been the work of examining things at the molecular level by purifying individual cellular components, especially proteins. The main method used for this has been chromatography.

Classical Types of Chromatography

The word chromatography means something like "color writing", a description that references early work in which the differently colored pigments in plant extracts were separated.

Let's describe chromatography in a single sentence. Different components of a mixture can be separated from each other if they partition differently between the two "phases" or elements of a chromatographic system, one of which is stationary (solid) and the other of which is mobile (liquid).

Many specific methods conform to this general definition, and represent variants of classical chromatography, even though the specific methods and materials have improved massively over the years. Non-mathematically and intuitively, this chapter considers, the common principles behind them as well as some of the specifics of each. You can read more sophisticated descriptions elsewhere when you have grasped the basics.

In protein and peptide science, the most important methods are ion-exchange chromatography, hydrophobic interaction chromatography, size-exclusion chromatography and reversed-phase chromatography. Table 12-1 summarizes these methods. Except for reversed-phase chromatography, which requires organic solvents, these methods all use hydrophilic column packing media and aqueous solvents so that soluble proteins can often retain their native structures and activities. They stand apart from the wider range of options applicable to small molecules, whether for preparative chemistry or purely analytical work.

Affinity Chromatography

Modern protein purification almost always begins with one or more steps of *affinity chromatography*, in which the solid phase selectively captures an individual protein bearing an engineered peptide tag or other readily captured group, such as biotin. For example, the column may carry an antibody that recognizes a certain peptide sequence or a chelated divalent metal ion (often Ni^{2+} or Co^{2+}) that attracts a hexahistidine tag built into a recombinant protein. The classical methods dealt with in this chapter are often used to remove residual contaminants from proteins after a first affinity-based step. Affinity chromatography is reviewed briefly in Chapter 13, but the rest of this chapter is restricted to conventional methods.

Table 12-1. Four Variants of Classical Chromatography Used with Proteins and Peptides

Chromatography	Stationary phase	Equilibration/loading	Eluant (mobile phase)	Driver of elution
Ion-exchange	Beads carrying positively or negatively charged groups on surface	Equilibrate column in aqueous buffer with low salt concentration: load sample in same	Aqueous buffer containing increasing gradient of salt, often NaCl	Proteins bind to column by ion-pairing of opposing charges: increasing concentrations of soluble ions in salt lead to elution
Hydrophobic interaction	Beads carrying neutral aromatic group or aliphatic chain	Equilibrate column in aqueous buffer with high salt concentration: load sample in same	Aqueous buffer containing decreasing gradient of salt, often $(NH_4)_2SO_4$ or NaCl	Proteins are initially "salted out" onto column: with decreasing salt concentrations, they enter the mobile phase and are eluted
Size-exclusion	Beads containing pores that differentially admit smaller molecules of the set to be separated while excluding larger ones	Equilibrate column in aqueous buffer with low salt concentration: load sample in same	Aqueous buffer containing low salt concentration to minimize nonspecific interactions of proteins with column: no gradient, isocratic elution	Pores in the packing medium tend to exclude the larger molecules in the sample while admitting the smaller ones: continuous isocratic elution results in fractionation
Reversed phase	Silica beads coated with aliphatic hydrocarbon chains (C_{18}, C_4) or beads composed of polymer	Equilibrate column in 0.1% formic acid or 0.1% trifluoroacetic acid: load sample in same	Increasing gradient of organic solvent, often acetonitrile (CH_3CN)	When mobile phase is nearly 100% aqueous, peptides partition onto stationary phase: with increasing concentrations of organic solvent, they are eluted

Separating Funnels – Solutes Partitioned Between Two Phases

Chromatography is best understood as an extension of a simpler type of separation. In the organic chemistry lab, most of us have seen how different compounds can partition differently between two phases such as an aqueous solution of chosen pH and an immiscible organic solvent such as ethyl acetate. When we shake our separating funnel and the two phases separate, compounds often favor one phase or the other depending on their polarity and charge. If our desired product enters one phase while unwanted impurities prefer the other, we have a useful separation.

Figure 12-1 illustrates a simple case in which two components of a green mixture are resolved by one step of partitioning. The blue component migrates 100% to the aqueous layer and the organic-soluble yellow component goes to the ethyl acetate layer.

In practice, we often need to repeat the separation to complete it.

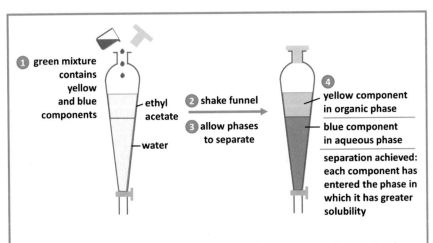

Figure 12-1. A separating funnel loaded with aqueous and organic phases provides a single partitioning experiment in which solutes enter one phase or the other depending on their respective solubilities. The green mixture (dissolved/suspended in water) contains a blue component and a yellow component, each of which partitions decisively into one of the two phases.

If our wanted material strongly prefers one of the two phases, we retain that phase and throw away the other. Then we introduce a fresh charge of the discarded phase and repeat the extraction.

Why Chromatography Is Like Calculus

The analogy between a separating funnel and chromatography is instructive, but so are their differences.

In the separating funnel, both phases are stationary. When we shake the funnel and two phases separate, each analyte is distributed between the two phases according to its solubility in each phase. Components that migrate into different phases are separated by this single round of partitioning, and the procedure serves chemists well when they separate an organic-soluble product from water-soluble impurities. Clearly, though, it is not equal to the challenge of separating multiple closely similar components from each other, as we must often do in biochemistry.

In chromatography, one phase is stationary and the other is mobile. Instead of a separation that takes place in discrete steps, we have one that is continuous.

Calculus deals with a continuous process like the acceleration of a car by breaking it down into tiny increments of movement, each of which occurs at its own constant speed. Chromatography conducts separations by allowing components of a mixture to "choose" or partition between two phases on a seemingly continuous basis over an extended period. Just as in calculus, we can imagine this process as consisting of a large number of individual partitioning experiments that are formally called "theoretical plates." Even if two components differ only a little in their relative preferences for the two phases, exposing them to a large number of theoretical plates can lead to those components being separated.

What Happens in Chromatography

The basic set-up is always the same. A liquid sample is loaded into a column that contains a packing called the stationary phase, and liquid called the mobile phase or eluant is pumped through.

- if the composition of the mobile phase never changes, chromatography is *isocratic* (from Greek roots: iso- means "equal", -cratic refers to power or strength)

- if the composition of the mobile phase changes progressively through the course of the experiment, with a steady increase or decrease in the concentration of salt or an organic solvent, this is *gradient* chromatography

Except for size-exclusion chromatography, in which analytes may reversibly migrate into the column packing but do not bind to it, initial conditions allow the components to be purified to bind to the stationary phase when first loaded. The initial conditions and the solvent for the sample must allow this, and are usually the same.

Then, if things go well, individual components of the sample are eluted as distinct populations or peaks at different stages of the overall process. By collecting the eluate (liquid flowing out of the column) in fractions, the separation is preserved so that further work can be done with the purified or part-purified components.

We like peaks to be "narrow" (eluted in a small fraction of the total time or volume of eluant, as in Figure 12-2A), because this gives the experiment "high resolution." This is the ability to resolve different components of the sample from each other. If the resolution is "low" or "poor", then the resulting wide peaks are more likely to overlap each other (Figure 12-2B). Where they overlap, components from each peak are present. Separation of the two components will be incomplete.

A 1 mg sample of a 20 kDa protein contains 50 nmol or 3×10^{16} protein molecules. Each one of this colossal number of molecules behaves independently, yet, when chromatography works well, they all emerge in a narrow band that we call a peak. This can only happen if they all have very nearly the same experience when they enter the column and interact with the two phases.

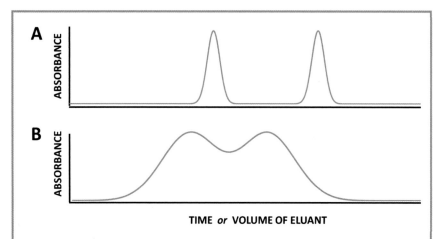

Figure 12-2. Consequences of (A) high and (B) low resolution in chromatography. The narrow peaks obtained in (A) allow the respective components in the two peaks to be fully separated. The wider peaks obtained in (B) cause separation to be incomplete.

At any instant – think of the separating funnel – they can either bind to the stationary phase or enter the mobile phase. In the former case, they make no progress toward the bottom of the column. In the latter case, they move along and are eluted unless they bind back to the packing.

Narrow peaks signal that the behavior of all the molecules of each type is highly consistent. This can be called "high performance" chromatography, and is the objective of all the effort devoted to developing superior materials and methods for separation. When the peaks are broad and resolution is poor, some factor is at work that is causing identical molecules to interact with the column in ways that give a divergence of results. This could be caused by inhomogeneous stationary phase (see next section), or even by poor column packing that caused molecules to encounter channels in the bed of chromatography medium.

Homogeneity and Particle Size of Stationary Phase Material

A crucial aspect of good chromatography is the homogeneity of the stationary phase material – the extent to which all the particles of the column packing are consistent with each other in size, porosity, hydrophobicity and density of active groups.

For all the molecules of a given sample component to have the same experience, all the beads of the packing material must be as close to identical as possible. Decades of progress have improved this aspect of all sorts of chromatographic media. After first being based on cellulose fibers, most media for protein purification are now based on beads of agarose or synthetic polymers. Beads for reversed-phase chromatography are formed from coated silica or synthetic polymers.

The physical basis of the separation should be considered. We are pumping liquid through a column tightly packed with beads. There is resistance to the flow of liquid, and it has to make its way through the gaps between the particles. If the resistance is too great, the pumps in our chromatography system will not be able to drive it.

Think of a large concrete sewer pipe that we might see at a construction site. If we pack it with basketballs (Figure 12-3A), then pump liquid through it, the gaps left between the large spheres will themselves be pretty big and liquid will be able to move through without much trouble. (We assume that the pipe is not too long, the balls are held in by a screen, and the pump can handle the task). If we packed the same pipe with golf balls, which have a five-fold smaller diameter (Figure 12-3B), more of its volume would be occupied by the packing and it would be much harder to drive liquid through at the same rate of flow.

Despite this difficulty, there is a major benefit attached to using the smallest possible particles for chromatography. A lot of the interaction between analytes and media takes place on the surface of the beads. As the surface area of a sphere increases with the square of its radius r, being given by the expression $4\pi r^2$, the amount of particle surface area present in a given volume of packing material is inversely proportional to the size of the particles. We can restate this in ordinary language – the smaller the beads in a fixed volume of packing, the greater the surface area that they present. In our improbable sewer pipe, the total surface area of the golf balls will greatly exceed that of the basketballs.

Analyte molecules are much smaller than beads of chromatographic media: a medium-sized globular protein might

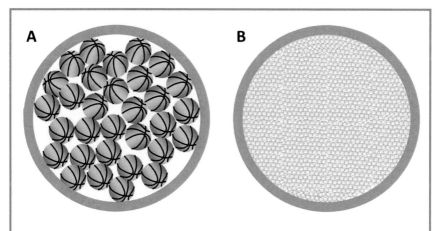

Figure 12-3. A pipe packed with (A) basketballs or (B) golf balls. There is a five-fold difference in the diameters of the two types of sphere.

have a diameter of 5 nm, while the smallest beads used in chromatography are 400 times wider (2 µm, or 2000 nm), and chromatographic media for proteins are generally even larger (Figure 12-4). Peptides – much shorter chains of linked amino-acid residues – are even smaller than proteins.

In addition to the nature of the packing material, the dimensions of the column are crucial to achieving the best performance. We often monitor the absorbance at an appropriate wavelength of the liquid exiting our column to detect and record the elution of analytes. It is the rise and fall of this signal that we record as peaks. The sensitivity of our peak detection depends on the concentration of the eluted analytes as they emerge – the higher the concentration, the greater the absorbance of the solution at the top of the peak, and the more easily we can detect it. One way to drive up the concentration is to achieve narrow peaks, which we discussed above. The other is to minimize the volume of liquid in which a peak emerges.

Reversed-phase high pressure liquid chromatography (RP-HPLC) is the mode of chromatography in which this principle is most easily seen. As pumps, media and methods have evolved, there has been a decided trend in favor of smaller media packed into progressively narrower columns and use of lower flow rates of solvent. This has not only provided more sensitive chromatography, but has also reduced the quantity of organic solvent used, which is environmentally favorable.

Overall, chromatography has evolved into a massive range of techniques for analytical and preparative purposes. A feel for its underlying principles turns a process that only a wizard can understand into one that always requires care and thought but can now be approached rationally.

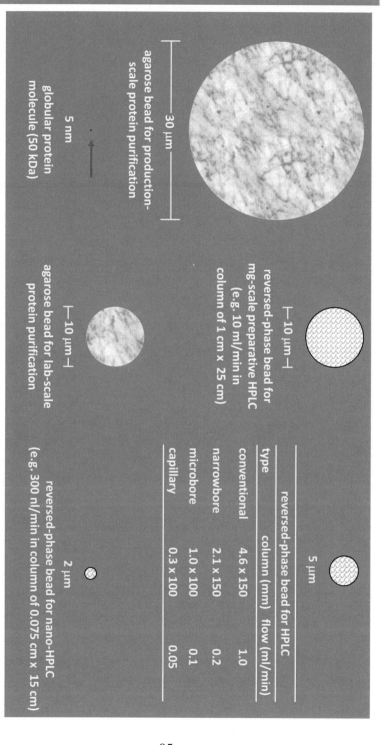

5 μm

reversed-phase bead for HPLC

type	column (mm)	flow (ml/min)
conventional	4.6 x 150	1.0
narrowbore	2.1 x 150	0.2
microbore	1.0 x 100	0.1
capillary	0.3 x 100	0.05

⊢ 10 μm ⊣
reversed-phase bead for
mg-scale preparative HPLC
(e.g. 10 ml/min in
column of 1 cm x 25 cm)

⊢ 30 μm ⊣
agarose bead for production-
scale protein purification

5 nm
globular protein
molecule (50 kDa)

⊢ 10 μm ⊣
agarose bead for lab-scale
protein purification

2 μm
reversed-phase bead for nano-HPLC
(e.g. 300 nl/min in column of 0.075 cm x 15 cm)

Figure 12-4. Cartoon depiction of selected chromatographic media with some representative information about column dimensions and flow rates used with them. Extensive variation is possible.

95

Affinity Chromatography

13

Conventional chromatography achieves purifications by exploiting differences between proteins that can often be quite subtle. This required biochemists from the 1950's into the 1980's to perform tedious multi-step procedures that often came with the added pleasure of needing to be performed in a cold room.

During these chilly sessions, the thought must have crossed many minds that there has to be a better way. Could there not be an ideal complementarity between the particular protein being purified and the stationary phase in chromatography? If that were so, protein purification would look much more like the decisive separations of hydrophilic and hydrophobic small molecules that are sometimes achieved in a separating funnel.

Affinity Chromatography Using Innate Properties of Proteins

Entering the 1960's, biochemists were convinced of the truth of the lock-and-key explanation for the binding of small molecules to large molecules. A 1968 paper by Cuatrecasas, Wilchek and Anfinsen is regarded as the occasion on which this insight was adapted to enzyme purification. Their idea was to immobilize a competitive inhibitor of a certain enzyme to a protein-compatible stationary phase, and rely on affinity between the targeted protein and immobilized ligand to cause the wanted protein to be captured selectively from a cell or tissue extract. Other proteins lacking that affinity would be washed away, and the captured protein would be eluted by changing the pH, adding a high concentration of salt, or introducing the ligand in soluble form.

No Change in Principle from Conventional Chromatography

The extreme partitioning represented by successful affinity chromatography is qualitatively the same as the partitioning achieved by conventional methods. The protein that we want binds to the stationary phase, while other proteins do not bind (Figure 13-1). The *principle* of chromatography is not changed, but the ligand-specificity of the target protein creates an extreme difference between its behavior and that of other proteins.

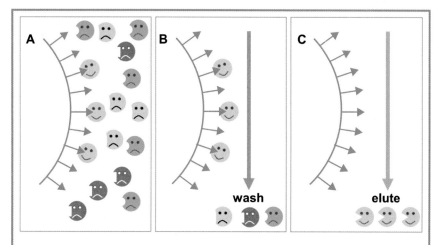

Figure 13-1. Idealized depiction of protein affinity chromatography. (A) In a crude extract, one protein (yellow smiley face) has affinity for an immobilized ligand (blue triangles). (B) Proteins that lack affinity for the immobilized ligand are washed away. (C) The purified protein is eluted.

Additional Approaches That Do Not Require Protein Engineering

The affinity of some proteins for a dye can also be made the basis of affinity purification. In particular, the ability of serum albumin (especially the human form) to bind to Cibacron Blue can be used to deplete plasma samples of this highly abundant protein prior to investigation of other proteins. The same dye provides a useful affinity matrix for capturing nucleotide-binding enzymes.

Recombinant DNA Changes the Game

In the 1980's, recombinant protein expression changed protein purification forever. Today, virtually every protein expressed for use in research is designed with its own purification in mind. There are also established methods for purification of antibodies based on proteins that naturally bind them.

Typically, the sequence of a recombinant protein includes an element designed to facilitate its purification from a cell extract or cell-conditioned medium. This may be a short peptide tag of around 20 amino-acid residues or less, or an independently folded protein domain that binds reversibly to an immobilized ligand.

It might seem that the hard task would be to find adequately tight-binding partnerships, but there is such a thing as *too much* affinity between partners. In particular, the otherwise valuable affinity between biotin and each of the proteins avidin and streptavidin is so tight, with K_d near 10^{-14} M, that it is not useful for affinity chromatography. Once a biotinylated protein binds to either of these proteins, strongly denaturing conditions are required to dissociate the complex. Fortunately, a modified form of avidin has been discovered that makes a useful replacement for the native protein.

Table 13-1 shows an incomplete list of widely used affinity chromatography systems. All can work well, and all have their complications. For example, a few proteins present in *Escherichia coli* cells exhibit high affinity for immobilized nickel or cobalt ions simply because their amino-acid sequences naturally contain pairs of adjacent histidine residues that clamp onto the metal chelate in the same way as a recombinant hexahistidine tag. Purifiers become accustomed to these recurring contaminants.

Table 13-1. Selected examples of affinity chromatography systems used in protein purification

Partner on stationary phase	Partner on protein	Eluant	Comment
Glutathione (linked to agarose through cysteine sulfur)	Glutathione S-transferase (GST) of *Schistosoma japonicum*	Glutathione	GST itself is dimeric, so GST fusions of dimeric proteins can be polymeric
Chelated divalent metal ion (often Ni^{2+} or Co^{2+})	Hexahistidine peptide tag	Imidazole	
Monomeric avidin	Biotin added in vivo or in vitro to 14-residue peptide tag	Biotin	Efficient biotinylation in vivo requires coexpression of the biotinyl transferase BirA
Starch (amylose)	Maltose binding protein	Maltose	
Anti-FLAG antibody	FLAG peptide, a particular short amino acid sequence added to the protein	FLAG peptide or pH 3.0 buffer	
Protein A or protein G	Constant region of immunoglobulin H chains (natural affinity)	pH 3.0 buffer	

The Fabulous Proteomic Algorithm

14

Fifty years ago, proteins were sequenced one amino-acid residue at a time, and every hard-won bit of sequence was a new discovery. Then protein sequencing led to molecular cloning, and cloning led to genomics, and faster, cheaper DNA sequencing allowed us to sequence the genome of any organism. From this we could predict quite accurately the amino-acid sequences of its proteins.

Period	1967-70's	1976-80's	1976-today	1993-today
Technology	Edman sequencing	molecular cloning	nucleic acid sequencing	proteomics
Scope	one protein	one gene	all genes	all proteins
Biological level	amino acid sequence	nucleic acid sequence	nucleic acid sequence	amino acid sequence

The result is that we never encounter a protein from a genome-sequenced species that is truly unknown to science. Whether the protein is apparently purified in a gel band or one of thousands in a cell extract, all we have to do is *recognize* it. Our task is to establish a one-to-one correspondence between the experimental material and a sequence entry in a database (Figure 14-1). In other words, we ask: *which* of the possible proteins belonging to the subject organism are we dealing with? We can answer that question because of the proteomic algorithm.

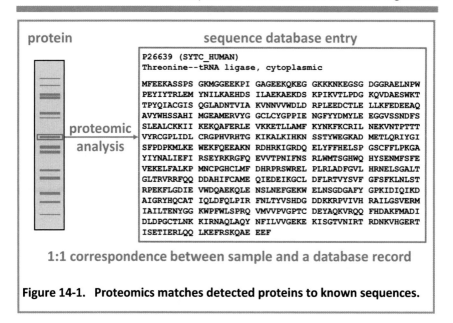

protein   sequence database entry

P26639 (SYTC_HUMAN)
Threonine--tRNA ligase, cytoplasmic

MFEEKASSPS GKMGGEEKPI GAGEEKQKEG GKKKNKEGSG DGGRAELNPW
PEYIYTRLEM YNILKAEHDS ILAEKAEKDS KPIKVTLPDG KQVDAESWKT
TPYQIACGIS QGLADNTVIA KVNNVVWDLD RPLEEDCTLE LLKFEDEEAQ
AVYWHSSAHI MGEAMERVYG GCLCYGPPIE NGFYYDMYLE EGGVSSNDFS
SLEALCKKII KEKQAFERLE VKKETLLAMF KYNKFKCRIL NEKVNTPTTT
VYRCGPLIDL CRGPHVRHTG KIKALKIHKN SSTYWEGKAD METLQRIYGI
SFPDPKMLKE WEKFQEEAKN RDHRKIGRDQ ELYFFHELSP GSCFFLPKGA
YIYNALIEFI RSEYRKRGFQ EVVTPNIFNS RLWMTSGHWQ HYSENMFSFE
VEKELFALKP MNCPGHCLMF DHRPRSWREL PLRLADFGVL HRNELSGALT
GLTRVRRFQQ DDAHIFCAME QIEDEIKGCL DFLRTVYSVF GFSFKLNLST
RPEKFLGDIE VWDQAEKQLE NSLNEFGEKW ELNSGDGAFY GPKIDIQIKD
AIGRYHQCAT IQLDFQLPIR FNLTYVSHDG DDKKRPVIVH RAILGSVERM
IAILTENYGG KWPFWLSPRQ VMVVPVGPTC DEYAQKVRQQ FHDAKFMADI
DLDPGCTLNK KIRNAQLAQY NFILVVGEKE KISGTVNIRT RDNKVHGERT
ISETIERLQQ LKEFRSKQAE EEF

1:1 correspondence between sample and a database record

Figure 14-1. Proteomics matches detected proteins to known sequences.

Origins of Proteomics

Genomics is the comprehensive study of a creature's genes. "Proteomics" is the study of the protein population of a biological system taken *en masse*. That system might not be a whole organism. We can speak of the human proteome – look up the Human Proteome Project if you like – but the proteomes of various human cells and tissues naturally differ as a result of local shifts in gene expression and protein post-translational modification. In differently specialized cells, a minority of the proteins support diverged functions while a larger and common set of "housekeeping" proteins maintain basic cellular operations.

Early proteomics (work done before the term was even coined) made use of two-dimensional gel electrophoresis (2D-GE) and Edman chemical sequencing, methods that by the end of the 1990's were supplanted by the rapidly evolving method of liquid chromatography-mass spectrometry (LC-MS).

In 2D-GE, electrophoretic fractionation of test (e.g. drug-treated or disease-related) and control samples gave patterns resembling the star patterns scanned by astronomers for changes in the night sky. Skygazers' software was even adapted for biochemists' use. Comparing patterns this way drew the experimenter's attention directly to proteins that changed in relative abundance or shifted in position due to modification when gel patterns were correlated, allowing analytical effort to focus on those proteins (Figure 14-2). Its main shortcomings were technical difficulty and detecting too few proteins to provide a deep dive into the proteome.

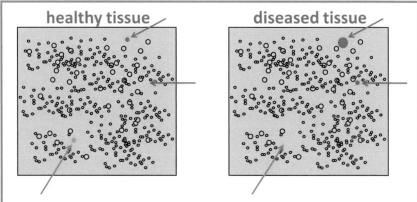

Figure 14-2. Two-dimensional gels directly highlight proteins that merit investigation, but detect many fewer proteins than LC-MS.

Mass Spectrometry Arrives

In this chapter, we will not attempt a full discussion of proteomics. Its subject is the pivotal algorithm that makes proteomics possible. **Key learning - Mass spectrometric analyses of peptides are correlated with information in protein sequence databases, leading to the identification of peptide fragments of known proteins.** After identifying peptides with high confidence, we can infer that their parent proteins were present.

Extensions of the process allow us to gauge changes in protein relative abundance between samples. Related terms like SILAC, iTRAQ® and TMT® are beyond this chapter's scope, but you can easily look them up and learn more.

Moving beyond the Edman method of chemical sequencing, the first mass spectrometry-based route to identifying proteins from 2D-GE gel spots was *peptide mass-mapping* or *mass-fingerprinting*. Debuting in 1993, this method combines the masses of multiple tryptic peptides into a list that is made the basis of a search: to emphasize, *multiple masses are bundled together to form one query*. Identity between several experimental masses and several theoretical peptides of a single protein yields an identification. As mass-mapping is in some ways a "quick-and-dirty" route to protein identification, MALDI-TOF mass spectrometry suits it well because it runs samples so quickly. Mass-mapping can be used to survey digests of a large number of relatively pure protein samples, but has been upstaged by the arrival of the fully-fledged proteomic algorithm and marvelous instruments that support its application.

This brilliant method makes a wizard or good witch of every protein scientist. In contrast to mass fingerprinting, which deals with collections of masses, each peptide analyzed is independently identified. Therefore, *many proteins can be digested as a mixture* and their peptides can be analyzed as a collection.

Evidence that a specific protein was present in a mixture comes from reviewing all the peptides identified. Detecting multiple peptide fragments of the same protein is persuasive evidence that their parent protein was among those digested. Even one peptide is cautiously believable evidence, although a protein identified on the basis of just one peptide is called a "one-hit wonder."

How It Works: Overview

Let's focus on how a single peptide in a mixture is identified. Like many computer-based strategies, proteomics handles complexity by repeating a single operation many times. The method applies to true proteomics, in which hundreds or thousands of proteins have to be recognized, but is also useful for simpler tasks, such as identifying the impurities in a protein preparation.

Assume we have a cell extract. When we add trypsin and incubate, the proteins fall apart into tryptic peptides. Thousands of proteins become tens of thousands of peptides. No mass spectrometer could detect and mass-analyze all the peptides at once, so we pull the mixture apart with a form of HPLC designed to elute each peptide in the absolute minimum volume so as to maximize its concentration and thereby favor its detection. Over time (often 1-3 hours), the eluate from a column trickles into a mass spectrometer. Any single peptide is delivered over a period set by the width of its chromatographic peak. With modern systems working well, this will often be near one minute. As thousands of peptides reach the spectrometer, each rising and falling in concentration over a different minute (and overlapping), we task the spectrometer with collecting the data that we need.

Figure 14-3 shows an analogy for the entire process, admittedly exotic. A collection of priceless vases bearing intricate patterns sits in a museum, with details of each one accurately recorded in a database of images. One day, a vandal smashes the vases into fragments. Examining each fragment with reference to the catalog allows it to be mapped to its parent vase. Reconstruction allows some of the value of the collection to be regained. *If we had no catalog, that would never have been possible.*

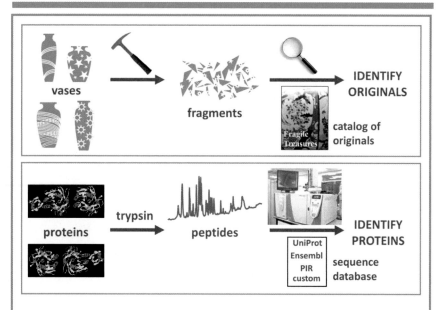

Figure 14-3. Proteomics by analogy. Upper panel: vases are smashed, then reconstructed by comparing their fragments with a catalog. Lower panel, proteins are fragmented, then identified by comparing their fragments (peptides) with a catalog (a sequence database, such as those listed).

Back to biochemistry. Proteins are fragile, complex biomolecules. After cutting them into simpler peptides, we identify the peptides using data from a mass spectrometer by reference to a sequence database. The results allow us to name the proteins that were present in the original sample. *If we had no database, that would never have been possible.*

So, if visiting aliens from the planet Zorgon kindly furnish you with samples of their blood to study, there will be little point in attempting proteomic analysis right away. Even if they assure you that their biology is based on DNA and L-amino acids, you don't have a reference database to underpin the study. Stick to your existing project until your guests have their genome(s) sequenced.

How It Works: Details

Chapter 10 showed how mass spectrometers are constructed to solve a series of problems. For proteomics (Figure 14-4), electrospray ionization and desorption (ESI) is always selected for the front end, while mass analysis is performed using a time-of-flight (TOF) analyzer or an Orbitrap® (a device exclusive to instruments from Thermo Fisher). The high mass accuracy of these devices is an important asset in proteomics, as we shall see below.

Again, let's consider one peptide out of the thousands in a complex sample. It is eluted from HPLC (often a nano-LC system running at a flow rate near 300 nl/min) in the form of a peak that is one minute wide or narrower. That gives us one minute or less to collect the information that will allow us to identify the peptide, and many peptides are competing for the attention of the instrument.

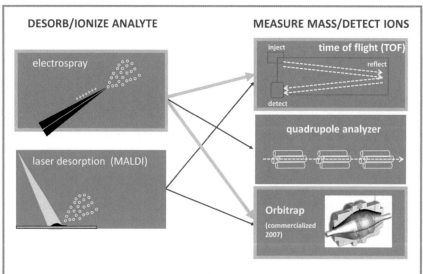

Figure 14-4. Mass spectrometers preferred for proteomics use electrospray ionization to send peptides into the gas phase, and either a time-of-flight or Orbitrap mass measuring device. The preferred options are framed in gold.

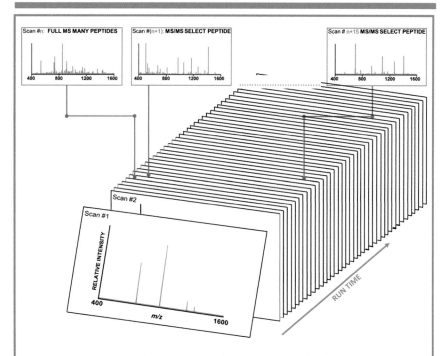

Figure 14-5. A proteomic dataset consists of a series of discrete mass spectra. More than one type of spectrum is collected, Peptides detected and mass-measured in "Full MS" scans are subsequently isolated from the continuing incoming supply and fragmented, after which mass spectra of the fragment sets are collected (MS/MS).

Key learning - The mass spectrometer collects its data as a series of discrete scans, but does not execute the same type of scan every time. When the run is complete, the data resemble an array of traditional library index cards (Figure 14-5), each card showing one mass spectrum. A modern instrument produces well over 20,000 scans in a 1-2 hour run, often allowing it to examine a high fraction of all the peptides in a complex sample.

Figure 14-5 shows how each mass spectrum is a discrete snapshot. This is quite different from the continuous signal collected from an absorbance detector attached to an HPLC.

The primary type of scan in proteomics measures the masses of intact peptides. It can be called "Full MS." ESI produces multiply-protonated and multiply-charged ions (Chapter 10), and peptides with masses of 600-2800 Da will give only a few peaks in a typical m/z range of 400-1600. Besides, as Figure 14-6B shows, a typically complex mixture of different peptides makes it impossible for us to apply the method shown for a single protein in Figure 10-5.

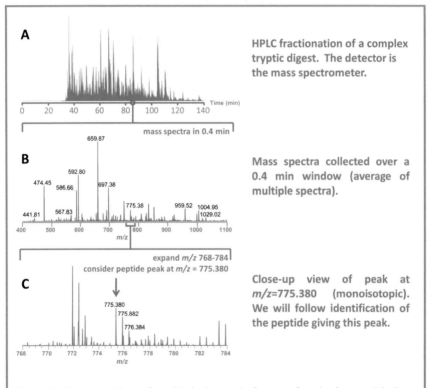

A HPLC fractionation of a complex tryptic digest. The detector is the mass spectrometer.

B Mass spectra collected over a 0.4 min window (average of multiple spectra).

C Close-up view of peak at m/z=775.380 (monoisotopic). We will follow identification of the peptide giving this peak.

Figure 14-6. Detection of multiple isotopic forms of a single peptide by a high-resolution mass spectrometer. (A) Chromatographic fractionation or "peptide map" of a tryptic digest of a cell lysate recorded by the mass spectrometer. (B) Full MS spectra collected over a 0.4 min section of the chromatogram and averaged. (C) Zoomed-in view of a peak cluster with m/z (monoisotopic) at 775.380. See the text for further discussion.

The fast and highly accurate mass spectrometers used in proteomics allow us to survey distinct features of crowded mass spectra independently. For example, even the weak peak cluster with monoisotopic $m/z=775.380$ (Figure 14-6C) may be selected for investigation, as the spectrometer is programmed to survey as many spectral features as possible rather that repeatedly concentrating on the strongest peaks.

It might be guessed that all molecules of a single peptide have the same mass, but we noted in Chapter 10 that 1.1% of carbon atoms in proteins and peptides are ^{13}C rather than the majority isotope ^{12}C. One of the attributes that make you a witty and charming individual is that about 1% of your own carbon atoms are ^{13}C.

Don't panic. ^{13}C is a *stable* isotope of carbon, and is not radioactive. Molecules in which some or all of the carbons are isotopically substituted with ^{13}C are widely used as internal standards in analytical mass spectrometry, including proteomics. An additional option is to introduce ^{15}N, another stable isotope, in place of the common ^{14}N. These standards are chemically identical to the analyte being studied, but differ in mass and can be distinguished from the corresponding analytes by a mass spectrometer. Adding them to samples at a known, fixed level provides an ideal means of quantitating variations in the concentrations of analytes. For some clinical (medical) purposes, it is even possible to infuse a stable-label bearing nutrient (e.g. an amino acid) into a patient to observe the rate of a metabolic process.

^{13}C differs from ^{12}C by having one extra neutron in its nucleus, and is 1 Da (actually 1.007 Da) heavier than ^{12}C. Peptide molecules in which one carbon atom is ^{13}C while the rest are ^{12}C are 1 Da heavier than those containing only ^{12}C (which are said to be *monoisotopic*). With two ^{13}C present, the molecule becomes 2 Da heavier.

As a result, mass analysis of a chemically pure peptide performed with a high-resolution instrument shows a cluster of several peaks (Figures 10-3B and 14-6C), each corresponding to a molecule with a different number of ^{13}C atoms in its complement of carbon. Any single carbon position has a 1% chance of being filled by ^{13}C, and a fifteen-residue peptide might have around 80 carbons. In a pure tryptic peptide, it is perfectly normal to find molecules containing different numbers of ^{13}C.

Figure 14-6C shows a peak cluster for a tryptic fragment of a human protein in which the monoisotopic peak is at $m/z = 775.380$, and the other peaks have m/z values that successively increase by 0.5. As molecules in each peak differ with respect to the number of ^{13}C atoms that they contain, the value of z for these peaks must be 2 because each replacement of ^{12}C by ^{13}C adds 1 Da to the mass.

Therefore, these related peaks represent isotopic variants of $[M+2H]^{2+}$ for a peptide fragment of a human protein that we wish to identify. Each of the two proton adducts contributes 1.007 Da to the mass of $[M+2H]^{2+}$, so the monoisotopic mass m of the peptide itself is found by the simple calculation:

$$\frac{m + 2.014}{2} = 775.380$$

$$m = 1548.746$$

If we consider that this measurement is accurate to within 10 parts per million of the true result, which is routinely achieved on a modern instrument, we can give the mass as 1548.746 ± 0.015 Da.

Therefore, the true mass lies in the range 1548.731-1548.761 Da.

Having measured the peptide mass rather accurately, we might ask if this measurement alone identifies a unique tryptic peptide from the human proteome. It does not!! Our search engine finds ten perfect tryptic fragments of human proteins with monoisotopic masses inside this window (Figure 14-7). Even if our peptide is one of these (and that is not yet certain), we do not know which one.

Mr(calc)	Δ(Da)	Sequence	Parent protein
1548.7460	0.0010	FLGDIEVWDQAEK	Threonine-tRNA ligase, cytoplasmic
1548.7355	0.0116	FISRNQEGPGEMGK	N-acetylgalactosaminyltransferase 13
1548.7453	0.0017	IDLDSMENSERIK	Centromere protein H
1548.7341	0.0129	DAVDSLGEAVDMSIK	BLOC-1-related complex subunit 6
1548.7503	-0.0032	MLKFLMFDMGLR	TBC1 domain family member 1
1548.7494	-0.0023	FLVMNDEGPVAETK	Uroplakin-3b-like proteins 1 and 2
1548.7541	-0.0071	MGIHFSCIRGDLK	Putative unchar. protein LOC388882
1548.7494	-0.0023	FLQVLPACTEDEK	Rotatin
1548.7532	-0.0062	EPQTTVIHNPDGNK	Calcium/CaM-dep. protein kinase IIdE
1548.7528	-0.0057	MTPPTTKPVMTDSK	Kinase suppressor of Ras 1

Figure 14-7. Tryptic fragments of human proteins with masses in the range of 1548.731-1548.761 Da as observed in our experiment. We require sequence-related information before we can designate one (or none) of the candidates as the detected peptide.

This happens all the time. Searching with a peptide mass alone identifies *candidates* for the identity of the detected peptide, but more information is needed for it to be identified uniquely.

What distinguishes each candidate peptide from the others? *Its amino-acid sequence.* To distinguish between the candidates, we need sequence-related information. This is where we rely on the ability of the mass spectrometer to perform additional operations. Recall the index cards in Figure 14-5 – we collected more than one type of spectrum. *The second type gives us the sequence-related information that we need.*

The first type of spectrum (Full MS) recorded mass-related information for a packet of ions, with ions at many values of *m/z* detected in the same spectrum (Figure 14-6B).

To produce the second type, the instrument first selects ions in a narrow window surrounding a particular *m/z* and discards all others. Preferably, but by no means always, this means retaining ions corresponding to a single peptide. The retained ions are accelerated by electric field shifts into such violent collisions with an inert gas that they break up. A spectrum is then acquired that shows *m/z* values of the resulting fragments.

Chemical bonds are resilient, like springs, and energizing peptide molecules by crashing them into collisions makes their bonds stretch and twist. If sufficient energy is imparted, bonds break. A level of force is selected so that many peptide molecules break only at one of their weakest points, which are most often the peptide bonds that connect amino-acid residues.

When we enter the organic chemistry lab to convert starting material A to product B, we want all the molecules to behave the same way so that we obtain a 100% yield. That's not the case when we fragment peptides in the mass spectrometer. In a sample of 100 fmol of peptide – a low but measurable level – we have around sixty billion molecules. Let's go through them one by one and see what happens to them.

The first molecule collides with helium and, by chance, breaks at the most C-terminal peptide bond. The larger fragment generated contains all but the C-terminal amino-acid residue. Meanwhile, the second molecule chances to break at the peptide bond two residues from the C-terminus. A different fragment is generated.

And so it goes. Reviewing the fate of each molecule shows us that roughly even levels of fragmentation at all the different peptide bonds produces a family of products (Figure 14-8). The masses of fragments making up this family *when taken as a collection* provide a characteristic fingerprint of the peptide being fragmented.

Returning to our example, the search algorithm uses this principle to consider which candidate listed in Figure 14-7 would give the observed fragmentation pattern. It calculates how each candidate would fragment and scores the quality of the match to the experimental spectrum. The candidate peptide with the highest score is proposed as the identity of the experimental peptide.

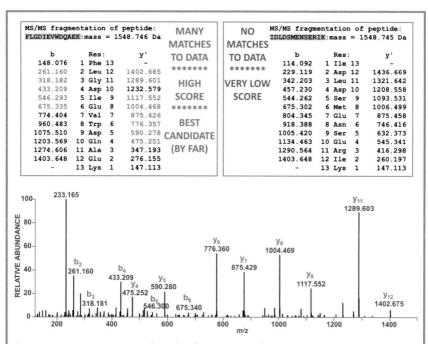

Figure 14-8. Sequence-related information allows the true identity of the 1548.746 Da peptide to be discovered. The tables at the top show predicted fragment ions expected from (left) the true hit and (right) one of the other candidates. Fragments detected in the experimental MS/MS spectrum are highlighted in red.

In our particular example, the top-scoring candidate is the one listed in the upper left corner of Figure 14-7. This peptide – FLGDIEVWDQAEK in one-letter amino-acid code – is a fragment of human threonine-tRNA ligase, and represents evidence that that protein was present in our original sample. Finding multiple peptides from this protein would make its presence virtually certain.

The proteomic algorithm is brilliantly successful, but we still must note that the top hit in a search of this kind is not always a true hit. For example, the experimental peptide might be a peptide that carries a post-translational modification such as phosphorylation, glycosylation or sulfation. These changes, of course, alter the mass of a peptide. Search programs can accommodate this possibility if it is factored into the search at the outset, but failure to anticipate a modification makes it impossible to identify a modified peptide correctly. This is because peptide mass is the first filter applied when we look for candidate identities in the sequence database, and chemical modification changes the mass of the peptide.

Proteomics is a massive and fast-developing subject, and this chapter has been limited to describing its very foundation. As the product of cross-fertilization between two consequential but unrelated scientific developments – large-scale DNA sequencing and protein/peptide mass spectrometry – it represents an object lesson in the unpredictable nature of scientific advance.

ACKNOWLEDGEMENTS AND SUGGESTIONS

The sage Lao Tzu teaches that a journey of a thousand miles begins with a single step. In science today, the beginner faces a long march from the foothills of learning into the high terrain of what is already known, but this road must be traveled if the frontier of new discovery is to be reached. So great is the volume of known material to be learned that some scientists receive an incomplete understanding of principles and technologies, especially those with a steep learning curve. It can be helpful to start our approach to a new topic with an easygoing explanation of basic aspects.

This book has attempted to provide gentle, intuitive and elementary explanations of its chosen topics. From the internet, textbooks and the scientific literature, virtually unlimited resources are available to enable readers to move on to more advanced learning, but the best preparation for dealing with advanced material is to have a secure grasp of the basics.

Chapter Notes

Chapters 1 and 2. Enzymology cannot really be made simple, but it is a pity to be repelled at the very beginning by its formalisms. The present book attempted to present fundamental aspects in an almost purely intuitive way, but serious engagement requires facing harder material. Two classic books in the field are Sir Alan Fersht's "Structure and Mechanism in Protein Science" and Christopher Walsh's "Enzymatic Reaction Mechanisms." Today, of course, the internet also offers vast quantities of relevant material, including class notes for college courses. None of these resources will be helpful unless the reader successfully relates the ever-present equations to what is really happening at the molecular level. Enabling that connection was our purpose.

Chapter 4. ATP – How the Cell Extracts Energy From Its Fuel. The main lesson offered in this chapter was one that I heard from the late Professor Paul B. Sigler of Yale University.

Chapters 2 and 4. Protein structural data were obtained from the RCSB Protein Data Bank (www.rcsb.org) (H.M. Berman, J. Westbrook, Z. Feng, G. Gilliland, T.N. Bhat, H. Weissig, I.N. Shindyalov, P.E. Bourne (2000) *Nucleic Acids Res.* 28: 235-242). Structural data set 4CHA is due to H. Tsukada and D. M. Blow (1985) *J. Mol. Biol.* 184: 703-711. Data set ISOQ is credited to J. A. Chamorro Gavilanes, J. A. Cuesta-Seijo, and S. Garcia-Granda (no reference cited). Data sets 4DW0 and 4DW1 were contributed by M. Hattori and E. Gouaux (2012) *Nature* 485: 207-212. The molecular graphics program RasMol was written by Roger Sayle.

Chapters 6 and 7. I learned something about optical spectroscopy from many colleagues over the years, especially Bart Holmquist, Alphonse Galdes, Rob Spencer and Boris Chrunyk.

Chapter 6. Molar absorptivity can be calculated from amino acid sequence using the online tool ProtParam found at the ExPASy web site maintained by the SIB Swiss Institute of Bioinformatics. The method implemented was originally described in S. C. Gill and P. H. von Hippel (1989) *Anal. Biochem.* 182, 319-326.

Chapter 8. The opening chapters of Stuart Kauffman's famous book "At Home in the Universe" offer accessible and fascinating lessons on the topic of how complex properties can emerge from repeated application of simple rules. Sequence diversity among aminoacyl tRNA synthetases is described in G. Eriani, M. Delarue, O. Poch, J. Gangloff and D. Moras (1990) *Nature* 347, 203–206.

Chapter 10 - Protein chemistry was transformed by the advent of mass spectrometry. I am very grateful to colleagues, especially

Justin Stroh, Tom McLellan and Xidong Feng, for their patient assistance in helping me to understand the potential of these amazing instruments. For a fine illustrated account of electrospray ionization, see M. Wilm (2011) *Mol. Cell. Proteomics* 10:M111.009407 (doi: 10.1074/mcp.M111.009407) and the tutorial PowerPoint file that accompanies it.

Chapter 11. Regarding SPR biosensors, I learned anything that I know from patient colleagues, especially Boris Chrunyk. For further reading, remember that most instrument vendors provide beautifully illustrated material explaining their products.

Chapter 13. The seminal paper in affinity chromatography referred to in this chapter is P. Cuatrecasas, M. Wilchek and C. B. Anfinsen (1968) *Proc. Natl. Acad. Sci. U.S.A.* 61, 636-643.

More broadly, I have sought enlightenment from so many scientists on so many topics over the years that they are too many to name. My gratitude and appreciation will never dim.

I greatly appreciated the help and patience of colleagues and friends who cast a puzzled eye over drafts of this eccentric book. They include Paul Engel, Chris Gabel, Lise Hoth, Alison Varghese, Alison Hackett and Graham West. Errors that remain despite their guidance are my own.

Kieran Geoghegan

2018

ABOUT THE AUTHOR

Kieran Geoghegan is originally from Dublin, Ireland, where he graduated in Biochemistry from University College, Dublin. He received his Ph.D. from Cambridge University, and was subsequently a post-doctoral fellow at the University of California, Davis and Harvard Medical School. He worked in drug discovery at a major pharmaceutical company in Connecticut from 1984 until his retirement in 2017, and has authored or coauthored more than 100 scientific publications. He lives in Mystic, Connecticut, and can be reached at kgeoghegan36@gmail.com.

Made in the USA
Middletown, DE
12 January 2020